Utility Security
The New Paradigm

Utility Security
The New Paradigm

Karl A. Seger, Ph.D.

PennWell Corporation
1421 South Sheridan Road
Tulsa, Oklahoma 74112-6600 USA

800.752.9764
+1.918.831.9421
sales@pennwell.com
www.pennwell-store.com
www.pennwell.com

Managing Editor: Kirk Bjornsgaard
Production Editor: Sue Rhodes Dodd
Cover design: Shanon Moore
Book design: John Potter

Library of Congress Cataloging-in-Publication Data

Seger, Karl A., 1940-
Utility security : the new paradigm / by Karl A. Seger.
p. cm.
Includes index.
ISBN 0-87814-882-5
1. Electric power plants--Security measures. 2. Computer networks--Security measures. 3. Public utilities--Security measures. 4. Terrorism--Prevention. I. Title.
TK1025 .S44 2003
333.793'2'0684--dc21

2003002795

Printed in the United States of America

1 2 3 4 5 07 06 05 04 03

To the victims of terrorism around the world
and to the valiant efforts of those fighting
this war on a day-to-day basis.

Contents

Figures

Tables

Acronyms

ALE	Annual Loss Expectancy
AMR	Automatic meter reading
ASD	Adversary sequence diagram
BATF	Bureau of Alcohol, Tobacco, and Firearms
CCTV	Closet circuit television
CDC	Center for Disease Control
CERT	Computer Emergency Response Team
CIPWG	Critical Infrastructure Protection Working Group
CSIRT	Computer Security Incident Response Teams
DOD	Department of Defense
DOE	Department of Energy
DOT	Department of Transportation
EAP	Employee Assistance Program
EASI	Estimate of adversary sequence interruption
ELF	Earth Liberation Front
EOC	Emergency Operations Center
EPA	Environmental Protection Agency
ES	Electric sector
ES-ISAC	Electricity Sector Information Sharing and Analysis Center
FAA	Federal Aviation Administration
FBI	Federal Bureau of Investigation
FEMA	Federal Emergency Management Agency
FVDS	Facility Vulnerability Determining System
HAZ-MAT	Hazardous material
IATF	Information Assurance Task Force
ISAC	Information Sharing and Analysis Center
ISAlliance	Internet Security Alliance
ISP	Internet Service Provider
IT	Information technology

IURPA	International Utility Revenue Protection Association
LNG	Liquefied natural gas
NERC	North American Electric Reliability Council
NIPC	National Infrastructure Protection Center
NSA	National Security Agency
NSTAC	National Security Telecommunications Advisory Committee
NWLF	New World Liberation Front
OPSEC	Operations security
OSHA	Occupational Safety and Health Association
PDD 63	Presidential Directive 63
PG&E	Pacific Gas & Electric
PPS	Physical Protection System
PSM	Process Safety Management
RCMP	Royal Canadian Mounted Police
RTU	Remote terminal unit
SCADA	Supervisory Control and Data Acquisition
SEC	Securities and Exchange Commission
SLA	Symbionese Liberation Army
SLO	Single Loss Occurrence
SPLC	Southern Poverty Law Center
VA	Vulnerability assessments
WTC	World Trade Center

Introduction

The world changed on September 11, 2001, and it will never return to September 10th. The terrorism genie is out of the bottle and awareness of this threat will not go away. It is a threat that people must learn to deal with as they continue with their careers and their lives.

Immediately following these deadly attacks, the U.S. utility industry realized how vulnerable it was to extremists, common criminals, and other threats.

- An alert was issued to all water systems in the United States on September 12, 2001, that terrorists might target them.
- On November 21, 2001, the North American Electric Reliability Council (NERC) issued its Threat Alert Levels and Physical Response Guidelines.

Utilities began assessing their security and making improvements.

The purpose of this book is to help utilities evaluate their security needs and implement measures to address those needs. The measures implemented should be commensurate with the threat. As threat levels change, so should security measures. This is the NERC concept, and it is emphasized throughout this book.

The intent of this book is not to make the reader a security expert but rather to help you understand the risk management and security process. The first section of the book focuses on the current threat to utilities. Section 2 provides information on security countermeasures that can help you address those threats. The appendices include tools and checklists to help in developing and improving your security program.

Utility security is not an option in today's world. It is a necessity. Good luck with your program.

SECTION 1 New Paradigm Threats

1

September 11, 2001

Events Leading to the Attacks

Did the events of September 11, 2001, change the world, or did those attacks simply make the world aware that a change had already taken place several years earlier? In fact, this change was first recognized during the investigation into the bombing of the 1993 World Trade Center (WTC). That WTC bombing was carried out by a group of Islamic extremists representing several different terrorist organizations. They had first considered bombing a number of separate Jewish targets in Brooklyn. However, the leader of the group, Ramzi Ahmed Yousef, decided they would kill more Jews by bombing the WTC. The objective was to kill as many people as possible. Terrorism shifted from organized and structured groups such as those we were confronting in the 1980s to the fluid network structure epitomized by al-Qaida. Terrorism has become more deadly, and it is being conducted by groups that are more difficult to monitor and predict.

Following the attacks of September 11, 2001, President Bush declared war on terrorism. Our main adversary, Osama bin Laden, had already declared war on the United States in 1996 after the bombing of the Khobar towers in Saudi Arabia that resulted in the deaths of 19 Americans and injuries to hundreds of others. Following the Khobar bombing, Osama appeared in an interview on American television, declaring that:

> *We believe that the biggest thieves in the world are Americans, and the biggest terrorists are Americans. The only way for us to defend these assaults is by using similar means. We do not differentiate between...uniforms and civilians.*

Osama bin Laden had declared war on America. But America failed to realize that it was at war until September 11, 2001.

This was not a war to be fought in a distant land but one to be fought on American soil. Al-Qaida showed Americans that they are vulnerable at home and that international terrorists can strike at the heart of America with devastating results. *Homeland Defense* became the new buzzword as government and industry began examining the vulnerability of America's infrastructure and how it could be improved.

There were concerns regarding the vulnerability of the nation's infrastructure before the attacks on September 11th, of course. On May 22, 1998, President Clinton signed Presidential Directive 63 (PDD 63), establishing the Administration's policy on critical infrastructure protection. PDD 63 recognized that although the United States possesses both the world's strongest military and its largest economy, both of these are reliant upon critical infrastructures and cyber-based information systems. The critical infrastructures identified in PDD 63 included, but were not limited to:

- information and communications
- banking and finance
- water supply
- aviation
- highways
- mass transit
- pipelines

- rail and waterborne commerce
- public health services
- electric power
- oil and gas production and storage

PDD 63 established the National Infrastructure Protection Center (NIPC) and set a goal of ensuring the capability to protect infrastructures from intentional acts by 2003. The NIPC includes representatives from the Federal Bureau of Investigation (FBI), Department of Defense (DOD), U.S. Secret Service, Department of Energy (DOE), Department of Transportation (DOT), the intelligence community, and the private sector. (Many of these agencies are proposed to be included in the Homeland Defense agency.) The mission of the NIPC includes the prevention of attacks and the coordination of the federal government's response to attacks including mitigation, investigation, and monitoring reconstitution efforts.

Attacking infrastructure targets in the United States is not new. Consider the action of the New World Liberation Front (NWLF), a leftist extremist group that was active on the West Coast during the 1970s and early 1980s. The NWLF used techniques similar to the Weather Underground-bombings accompanied by warnings and demands. Their first bombing targeted an insurance agency in Burlingame, California, on August 5, 1974. In 1975, NWLF claimed credit for 22 bombings, most of them targeting Pacific Gas and Electric (PG&E) transmission towers:[1]

Feb 3 ***San Jose, CA*** An office building housing General Motors and Pacific Telephone Company is bombed.

Feb 4 ***El Granada, CA*** A bomb is placed near a fuel tank at a U.S. Air Force installation.

Feb 4 ***Oakland, CA*** The Oakland Asphalt Company is bombed.

Feb 5 ***Oakland, CA*** The Vulcan Foundry is bombed. The NWLF claims the Chevron Asphalt Company located next door was the intended target.

Feb 6 ***San Francisco,* CA** The fire exits at a television station are bombed. The NWLF telephones a warning to the station before the bombing.

Mar 12 ***San Bruno Hills,* CA** A bomb brings down a PG&E transmission tower.

Mar 20 ***San Bruno,* CA** Six transmission towers are bombed.

Mar 27 ***San Jose,* CA** Five bombs explode at PG&E targets. A letter from the NWLF demands free utilities for unemployed persons.

Mar 29 ***Sacramento,* CA** A PG&E transformer bank at the McDonnell-Douglas Company is bombed.

Apr 8 ***San Jose,* CA** A PG&E substation bombed in March is hit again. The NWLF now demands free electric service for retired people.

May 1 ***Sacramento,* CA** A California Department of Corrections office is bombed.

May 9 ***Berkeley,* CA** PG&E office is bombed. A caller demands free utilities for retired people and lower rates for the poor.

May 19 ***Tamal,* CA** A firearms range house outside of San Quentin is bombed.

Aug 8 **CA** Bombs are detonated at the residential estate of a member of Safeway's board of directors.

Aug 20 ***San Rafael,* CA** Two Marin County sheriff's patrol cars are bombed. The bombing is conducted along with members of the Symbionese Liberation Army (SLA), the group that kidnapped heiress Patty Hearst.

Sept 12 ***Seattle,* WA** A unit of the NWLF attempts to bomb the Veterans Administration office.

Sept 12 ***Phoenix,* AZ** Another NWLF unit places a bomb in a federal building.

Oct 13 ***Redwood City,* CA** An undetonated bomb is found next to a PG&E transmission tower.

Oct 31 ***Fort Ord,* CA** Bombing of a storage building. NWLF claims the bombing is in support of Puerto Rican nationalists.

Nov 30 ***San Francisco, CA*** A car belonging to the owner of a hospital equipment firm is bombed. The NWLF demands two free medical centers for the poor.

Dec 2 ***San Francisco, CA*** A bomb damages a Mercedes Benz parked in the driveway at the home of a dentist. The NWLF again demands two free medical clinics.

Dec 11 ***San Francisco, CA*** The NWLF publishes an open letter demanding $100,000 to improve medical care at the San Bruno County Jail.

Utility infrastructures continue to be targeted by extremists. In 1997, four Ku Klux Klan members pleaded guilty to conspiring to bomb a natural gas refinery near Bridgeport, Texas. The blast, which was to be a diversion while the group robbed an armored car, would have resulted in extensive damage and considerable loss of lives.

Members of the anti-government San Joaquin Militia have been convicted of attempting to bomb liquefied propane storage tanks in Elk Grove, California. The explosion could have resulted in thousands of casualties. The plot was designed to create civil unrest that would have resulted in martial law and the eventual fall of the government. The militia group had also planned to assassinate a local judge.

Two men were indicted for planning a Jihad-type mission in south Florida during 2001. Their targets were a National Guard armory and several Florida Power and Light substations in Dade and Broward counties. They were arrested while attempting to purchase AK-47 assault rifles, night vision equipment, stun guns, pepper spray, smoke grenades, and other military equipment.

Post September 11, 2001 Responses

As noted, the utility industry responded quickly to the events of September 11th. Security standards were reviewed and updated as needed by national associations representing the electric, gas, oil, and water/wastewater industries. As part of this response, the North American Electric Reliability Council (NERC), working with the Critical Infrastructure Protection Working Group (CIPWG) and the Edison Electric Institute Security Committee, published the *Threat Alert levels and Physical Response Guidelines* on November 26, 2001. The guidelines were revised (version 2.0) on October 8, 2002 to reflect the alert levels used by the Department of Homeland Security and the NIPC. They provide a model for developing organization-specific threat response plans using five alert levels.[2]

Threatcon ES-Physical-Green (Low) applies when no known threat exists of terrorist activity or only a general concern exists about criminal activity (such as vandalism), which warrants only routine security procedures. Any security measures applied should be maintainable indefinitely and without adverse impact to facility operations. This level is equivalent to normal daily operations.

Threatcon ES-Physical-Blue (Guarded) applies when a general threat exists of terrorist or increased criminal activity with no specific threat directed against the electric industry. Additional security measures are recommended, and they should be maintainable for an indefinite period of time with minimum impact on normal facility operations.

Threatcon ES-Physical-Yellow (Elevated) applies when a general threat exists of terrorist or criminal activity directed against the electric industry. Implementation of additional security measures is expected. Such measures are anticipated to last for an indefinite period of time.

Threatcon ES-Physical-Orange (High) applies when a credible threat exists of terrorist or criminal activity directed against the electric industry. Additional security measures have been implemented. Such measures may be anticipated to last for a defined period of time.

Threatcon ES-Physical-Red (Severe) applies when an incident occurs or credible intelligence information is received by the electric industry indicating a terrorist or criminal act against the electric industry is imminent or has occurred. This condition may apply as a result of an incident in North America outside of the electricity sector. Maximum security measures are necessary. Implementation of such measures could cause hardship on personnel and seriously impact facility business and security activities.

The NERC threat condition system is similar to those used by the U.S. Department of Defense (DOD) and the Federal Aviation Administration (FAA). During normal conditions, normal security measures should be in place. As the threat condition level increases, additional security measures are initiated. The NERC system is not limited to the electric industry and is appropriate for all utilities and indeed for all businesses and government organizations.

Threat condition levels can be increased based on threats other than criminal or terrorist—for example, in response to anticipated labor actions or threatened or actual natural disasters. In addition, the local utility doesn't need an alert from the Information Sharing and Analysis Center (ISAC) or the NIPC to increase the threat level. Local management can initiate a threat condition increase when a threat is perceived or imminent.

Every utility should have a NERC-type threat condition plan. Procedures for developing, implementing, and testing the plan at your utility are discussed in chapter 10.

While the September 11 attacks created an increased awareness in the utility industry for the need for ongoing security programs, it also raised the bar in terms of what is considered "normal." Many security measures that have been initiated since the attacks should have been in place before September 11, especially access control and physical security measures.

In some cases, utilities and other organizations may be overresponding to baseline security measures. There may be more security in place than is needed. As a risk management issue, security should always be commensurate with the level of threat, and plans must be in place to initiate additional security measures as the threat level increases. Potential threats must be constantly evaluated, and Threatcon levels increased or decreased as appropriate. These threats are not limited to terrorist or extremist activities and should be based on a broader workplace violence prevention model (Table 1–1). As noted in this model, the number one security threat to utilities is not a terrorist or extremist action. You are more likely to be targeted by a disgruntled ex-employee or an angry customer. You must be prepared to identify and respond to those threats that are likely to occur. The workplace violence prevention approach to threat assessments will be discussed in more detail in chapter 4.

Table 1–1 Workplace Violence Prevention Model

Category	Probability of Occurrence	Probable Consequence
Enraged Employee or Ex-Employee	High Probability	High Consequence Possible to Targeted Individual
Domestic Violence	Variable; Need to Monitor Employees with Domestic Problems	High Consequence Possible to Targeted Individual
Enraged Customer	Always a Potential Threat in the Utility Industry	Consequences Usually Controlled
Civil Dispute	Inevitable, e.g., Right-of-Way Disputes	Can Result in Highly Emotional and Dangerous Response
Robbery or Other Crime	Potential Threat Where Money or Other Valuable Assets are Present	Usually Results in Loss of Asset Targeted
Deranged Individual	Unpredictable	High Consequence Possible
Terrorist or Other Extremist Action	Low Probability	Highest Potential Consequence

The objective of this book is to provide an overview of the threats to utilities and information on how you can manage those threats. This is not a "one-size-fits-all" approach. The threat and corresponding security needs vary, depending upon your geographic location and the type of utility services you provide—and the threat is constantly changing. In Section 1, the general threats to infrastructure targets are discussed. In Section 2, you will learn how to develop a dynamic threat assessment capability for your utility.

Section 2 also instructs you how to conduct a vulnerability survey and how to develop a vulnerability-countermeasures matrix. Formulae for determining the cost-effectiveness of the measures considered are also presented. This section provides information on specific suggestions for physical security, access control, lighting, and other security measures. Two other important topics are explored—operations security and the security of your personnel.

Chapter 10 outlines the development of a NERC-type threat condition response plan. Other industries have learned that, in addition to a threat condition response plan, specific procedures and responsibilities are needed for implementing the plan and they must be tested.

The appendix to the book includes checklists, examples, and other information to help you implement the procedures discussed.

This is designed to be a handbook, a working document that will help you assess and hopefully improve the security at your utility. The examples provided are from actual incidents that occurred at utilities, although the location or name of the utility is omitted unless it was released by the utility or published in a major publication.

The paradigm for utility security shifted radically on September 11, 2001. Security is no longer an option; it is a necessity that has become a priority at most utilities. This book should help you make the paradigm shift and provide your utility with the security your employees and customers deserve.

Notes

1. *Disorders and Terrorism: Report of the Task Force on Disorders and Terrorism* (Washington, D.C.: National Advisory Committee on Criminal Justice Standards and Goals), 1976, pp. 20–21

2. *Threat Alert System and Physical Response Guidelines for the Electric Sector, Version 2.0*, North American Reliability Council Critical Infrastructure Protection Advisory Group, October 8, 2002

2

The Threat From Extremists

Kill one and frighten ten thousand.

— An old Chinese saying

Violence is the language that western democracies understand.

— Carlos the Jackal

The Terrorist Threat

Unless you are in a major city or near a highly sensitive facility, the probability of Osama bin Laden or the al-Qaida targeting your utility or community is extremely small. Still, the terrorist threat is real. A number of international terrorist groups have an operational capability within the United States, and countless domestic groups and potential local terrorists pose a threat to your utility. Since the ultimate infrastructure threat is terrorism, let's examine terrorist groups, how they are organized, how they function, and why they target utilities.

Terrorism is the practice of indiscriminate violence against random civilian populations. By its nature, it is faceless, ruthless, and addressless. The FBI defines terrorism as:

> *... the unlawful use of force or violence against persons or property to intimidate or coerce a government, the civilian population, or any segment thereof, in furtherance of political or social objectives.*

A significant problem confronting us in today's world is the impact of the media. Terrorists no longer have to be limited to frightening ten thousand; they can frighten millions. September 11, 2001, was witnessed by millions of people around the world. Those Americans who didn't actually see the collapse of the WTC towers saw television reports over and over and over. There is little doubt that the terrorists chose to go ahead with the operation when they did because it was a bright sunny day in New York and Washington, D.C.—a day when cameras couldn't help but record the attacks. If there had been poor visibility and the upper levels of the WTC could not have been seen because of clouds or fog, the terrorists would have likely waited for one of their fallback dates. There is evidence that the suicide hijackers had reservations on other flights after September 11, 2001.

To understand terrorism, you must examine the following elements of its true impact:

- The terrorist group and its objectives
- The terrorist incident and its impact on the immediate victims
- Media response and public reaction
- Government response of subsequent policy decisions

To understand how the dynamics of these criteria changed on September 11, 2001, let's first discuss incidents that occurred prior to that date.

On October 23, 1983, a suicide bomber drove a truck loaded with approximately 5000 pounds of explosives into the U.S. Marine Corps barracks at the Beirut airport—241 Marines died in the attack. Another suicide bomber drove a similar bomb into the French paratrooper barracks at almost exactly the same time—55 soldiers died in that attack. The group responsible for the attacks was the Lebanese Hizbollah, a Shiite extremist group sponsored by Iran and Syria. The devastation of the attack and the bodies of the dead and wounded were shown repeatedly on U.S. television. The objective of the attack was to convince the U.S. government to withdraw from Lebanon. Several months following the bombing, the Marines were withdrawn. The terrorists achieved their objective.

But the paradigm shifted on September 11, 2001. The objective of those attacks was to let America know that it was at war with Islamic extremists who insisted America remove its troops from the Saudi peninsula and stop supporting the state of Israel. However, al-Qaida totally underestimated the response to their atrocious actions. The al-Qaida is on the run (but certainly not out of business), and the government that supported it is history. The United States and much of the world are now committed to the war against terrorism. But the terrorists are equally committed to continuing the war.

Their underestimation of the response to the September 11th attacks was discussed in the New York Times on December 9, 2001.

> *In some ways, then, September 11 really was a suicide mission for all al-Qaida as a whole. It didn't just kill the terrorists involved; it sealed the fate of their supporters as well.*

Terrorist groups, whether international or domestic, have the same basic structure and operational characteristics. The groups include two levels of membership and are supported by active and passive supporters.[1]

At the top of the group are the leaders. These are usually deeply committed and highly intelligent individuals. Many have university degrees and some have advanced degrees. They are dynamic leaders and effective organizers. However, they rarely participate in the attacks. They function more like managers of a larger organization and let others do the fieldwork.

The group members that do the fieldwork are referred to as the active cadre. These people collect intelligence on potential targets, are responsible for logistics at the local level, and prepare for and conduct the operations. These are the soldiers of the organization.

The terrorist group cannot function without active and passive supporters. These are not members of the group but people who believe in the goals and objectives of the terrorist organization. Passive members will not commit illegal acts but may contribute money and other assets to the group. Active supporters will not set off the bombs or participate in other actions, but they will provide safe houses for group members, buy weapons for them, collect intelligence, or commit other illegal acts in support of the group and its goals.

There is an established set of criteria that terrorist groups use when selecting potential targets. It is important to note that groups rarely select and collect intelligence on a single target. They usually collect intelligence on a number of potential targets and then select the one that will most likely allow them to achieve the objectives of the attack. In the planning that led to the event of September 11, 2001, there is little doubt that the terrorists examined different airports and airlines to identify those flights where they would have the greatest probability of seizing the aircraft and flying them into the targeted buildings according to the designated schedule. And, as already discussed, we know that there were fallback dates and plans.

These are the criteria that terrorists use when selecting targets:

- Criticality of the target: If they hit it, will anyone know or care?
- Accessibility of the target: Can they get to it?
- Media impact: How much media coverage will the event receive?
- Vulnerability of the target: What will it take to achieve their objectives?
- Effect on the terrorist group: al-Qaida obviously underestimated this variable in planning the attacks on September 11th.
- Risk to the terrorist cell: This is obviously not a consideration when dealing with suicide terrorists.

What is the probability that a terrorist group or an extremist organization will target your utility? You can begin to assess the threat by applying the above criteria to the assets you are protecting. But remember that the threat to most local utilities in the United States comes from *domestic* groups and *individual* terrorists. They use the same target criteria but approach targeting decisions from a different perspective. They may be focused on regional or local issues, and disrupting utility services in your area may help them to gain the attention for the cause to which they are committed.

The structure of both international and domestic terrorist groups changed radically in the early 1990s. Domestic terrorist groups across North America adopted a *leaderless warfare* concept. They no longer form organizations of 20 to 30 members but function in cells of 2 to 4 people. These cells keep to themselves, which makes it more difficult for law enforcement to identify and investigate them. The ultimate terrorist was the Unabomber, a single terrorist who might still be killing people if his family had not reported their suspicions to the FBI.

International groups began to adopt an *all-channel* network structure prior to the 1993 WTC bombing. The network structure allows members of different organizations to operate together and to continue operations even if one of the cells or nodes in the network is disrupted. This is the unfortunate genius of al-Qaida. It is not really one structured organization but a number of different cells. In addition, its cells operate in concert with other terrorist organizations such as Hizbollah and the various Islamic Jihad groups. For example, the networks and groups participating in the 1993 WTC bombing included:

- Gama'at'Ilamiyya, Egypt
- National Islamic Front, Sudan
- Hamas, Israel
- Islamic Jihad, Lebanon
- Ul-Fuqra, a group originally organized in Colorado

To summarize, the terrorist threat changed radically during the 1990s. Terrorist groups today are much more fluid; they are based on an "all-channel" network model of organization, and they are committed to increasingly deadly attacks. This includes both international terrorists and domestic groups. Although al-Qaida was responsible for the approximately 3000 deaths on September 11, 2001, homegrown terrorists committed the bombing of the Murrah building in Oklahoma City in 1995 and planned bombings at a natural gas refinery in Texas in 1997 and at a liquefied propane storage facility in California in 2000. Timothy McVeigh and the other terrorists associated with the Oklahoma City bombing intentionally murdered innocent children. If carried out, the bombings in Texas and California could have killed more people than those who died on September 11, 2001.

Domestic threat

In the middle 1980s, a U.S. electric utility caught a religious group living in a commune stealing energy. The leaders of the group apologized to the utility, claimed they didn't know what they were doing was illegal, and they paid the estimated back bill. Several months later, federal agents raiding the compound found that the group had been manufacturing automatic weapons for white supremacist groups across the United States and were making homemade land mines to protect the commune. At least one of the members arrested during the raid was subsequently convicted for the murder of a police officer. This group decided in 1983 to bomb the federal building in Oklahoma City but canceled the plan when part of the bomb exploded as it was being built. They also had plans to bomb electric transmission lines and to poison water supplies. The commune member who was arrested for the murder of the police officer, Richard Snell, was executed in Arkansas the night before the 1995 Oklahoma City bombing.

In 1997, four Ku Klux Klan members were convicted for conspiring to bomb hydrogen sulfide tanks at a natural gas refinery in Wise County, Texas. They had already purchased the materials needed to build the bombs, and authorities report they had the expertise needed to carry out the plan. The bombers planned to maximize casualties by staggering the timing on the explosions so that the main charges would go off *after* law enforcement and other first responders arrived on the scene. The explosions, which could have destroyed half of the county, were planned as a diversion while the group robbed an armored car.

To help ring in the Y2K New Year, a group of militia members in California planned to blow up the largest liquefied propane storage tanks in the United States. The storage tanks are located near an interstate highway and a housing subdivision where the kill radius would have been five miles. An estimated 2000 people would have died. The group's objective was to start a civil war that they believed would have caused the U.S. government to fall.

Utilities always make tempting targets for extremists. In the 1970s, a neo-Nazi group in the midwestern United States planned to poison a local water supply of a predominately Jewish community. A *skinhead* group planned a similar attack in Texas in the 1980s.

As the NWLF demonstrated during the 1970s and 1980s, leftist groups also target utilities. When the Pope visited Peru, a major left-wing terrorist group bombed transmission towers and put out the lights in Lima. During the height of the Red Army Faction activities in Germany, leftists bombed almost 150 utility lines in a single year.

Special interest groups and organized crime also pose a threat to utilities. In one case, a utility found and prosecuted power diversion at a business owned by a local organized crime group. A subsequent arson fire occurred at one of the utility's field offices. In a similar case, an organized crime group, caught stealing utility services at their business, torched the rental property of the utility employee who caught them.

One of the special interest groups that represent a threat to utilities is the Earth Liberation Front (ELF). Their most infamous act was a $12 million arson of a ski resort in Vail, Colorado. ELF developed a new type of incendiary device for the attacks (the plans for the devices are described on their web site), and they set off a number of the devices simultaneously. This group has been responsible for a number of arsons in other areas where they oppose *urban sprawl*, which they define as new housing being built in rural areas.

Tactics

Traditional tactics used by terrorists and other extremist groups include:

- bombs
- arson
- assassination
- hostage taking
- ambush and raids
- hijacking
- seizure of facilities
- sabotage
- civil disorder
- cyber attacks
- hoaxes

Tactics used against utilities have been primarily bombs, arson, sabotage, and hoaxes (*i.e.*, bomb threats). Most infrastructure protection professionals believe that we will begin to experience cyber attacks as well.

Bombs account for at least 50% of all terrorist attacks. Since the 1980s, there have been several significant trends in the use of bombs by terrorists. The most alarming of these is to maximize casualties. The intent of the 1993 bombing of the WTC was to kill thousands of people. The bombers expected the device to cause one of the towers to fall into the other, destroying both buildings. They also placed cyanide with the bomb thinking that the gas would be released into the towers upon detonation. Fortunately, the heat of the explosion caused the cyanide to evaporate without releasing any gas.[2]

Terrorist bombs often have anti-tampering components. For example, a mercury switch is placed in the device so if it is tilted or moved, the switch will complete the circuit and the bomb will explode. Anti-tampering systems are built into the device in addition to the primary means of detonation (often a timer).

To maximize casualties, extremists also use secondary devices. When anti-abortionist Eric Rudolph placed a bomb at a clinic in Birmingham, Alabama, he also placed a secondary device; it was timed to explode after police and other first responders arrived at the scene. The secondary device exploded, wounding a federal law enforcement officer who was in the process of investigating the first explosion.

Approximately 80 to 90% of improvised explosive devices (homemade bombs) are made using smokeless powder or dynamite. If Semtex, C4, or other more sophisticated materials are used, it may be an indication that the group is receiving support from a *state sponsor*.

Truck and car bombs present a special threat. A tremendous amount of explosive materials can be packed into a van or truck with devastating results. Some of these bombs (like the one used in 1993 at the WTC) require some degree of expertise, but others, such as the McVeigh truck bomb, are surprisingly simple in their design. In both incidents, the bombs were highly destructive.

Letter and package bombs are also a concern. Terrorist letter bombs usually run in cycles, *i.e.*, a group will decide to use letter/package bombs as a tactic and sends them to a number of different targets. Currently, the major threat to utilities is a mail/package bomb sent by an angry customer or disgruntled ex-employee; to date, extremist groups have not targeted utilities for letter or package bomb attacks.

On September 11, 2001, the United States was attacked by suicide terrorists who used commercial aircraft as their weapons. But the United States has not experienced the type of suicide bombers that have been active in Israel and other countries for many years. This is because a terrorist infrastructure must be present, such as the Hamas military wing in the Middle East or the Tamil Tigers in Sri Lanka, to support an extended suicide bombing campaign. It is not hard to find individuals in either country who want to become martyrs. They have a surprising number of volunteers from which to select. The group has to be able to identify these best candidates, indoctrinate and train them, and control them up to the point of carrying our their mission to make sure they do not change their minds at the last minute.[3]

The threat of suicide bombings is also present in countries where this group infrastructure does not exist, but the threat is limited. Domestic terrorists may decide to carry out a suicide attack on their own, or an international group may send *so-called* martyrs to the country to carry out a suicide mission. However, the frequency of the attacks will be limited where the terrorist group infrastructure does not exist to support them.

Bombs and explosive devices represent about 50% of all terrorist attacks. Another 15 to 20% are arsons.

Utility systems have to be concerned with the possibility of sabotage since it is impossible to protect all facilities required to deliver services to the end user. Power plants, electric substations, water storage tanks, pumping stations, and gas compression stations could be targeted; transmission lines and pipelines have been targeted by extremists and disgruntled customers. Extremists have also targeted propane and natural gas processing and storage facilities.

Once an extremist group has established its credibility by attacking targets, it can disrupt operations simply by threatening to detonate a bomb or some other action. If your utility receives a threat from a known militia or militant environmental group you *must* respond. Depending upon the nature of the threat, the utility may increase security at remote locations or even close the main offices and customer services areas. A bomb may or may not go off, sabotage may or may not occur, but your operations have been disrupted. That is the terrorists' intent!

There has been an emphasis in recent years to protect infrastructure systems from cyber attacks. This includes utilities. As the world becomes more automated and connected, this threat increases. Supervisory Control and Data Acquisition (SCADA) systems may be susceptible to hackers and other criminals; control systems at electric substations and other critical points could be compromised, resulting in damage to portions of the system. Although system vendors and the professionals who operate them make every effort to ensure that they are secure, compromises do take place. In April and May of 2001, during the height of the California energy crisis, hackers mounted an attack against the Cal-ISO computers that were critical in controlling the transmission of electricity around the state. The attacks began around April 25, but they were not detected until May 11. The hackers were able to enter the system from Internet servers located in Santa Clara, California, Tulsa, Oklahoma, and through China Telecom.[4]

Weapons of mass destruction

Weapons of mass destruction include nuclear and radiological bombs and chemical and biological weapons. While there have been attempts to manufacture and use these weapons, none of the known attacks targeted utilities. The most notable example in the United States was the poisoning of a number of salad bars in Oregon in 1983 by a cult trying to influence local elections![5]

The probability that an extremist group could obtain a nuclear device is almost nonexistent. The group would have to gain access to bomb-grade materials and have the manufacturing equipment and expertise to build the device. It's a long way from downloading plans for a nuclear device from the Internet to actually building a bomb. There is concern that a state sponsor could give a terrorist group a nuclear device, but this probability is also limited. Rogue states that have nuclear capabilities have only a few actual bombs, and if any were used by a terrorist group, they could be traced to their origin.

The media has speculated that terrorists may purchase a *suitcase bomb*, stolen from the Russian military. While there is concern that some of these weapons are missing, this threat also has a low probability; these tactical nuclear devices were considerably larger than a suitcase and designed to be deployed by a five-person special operations team, not a lone terrorist. In addition, some of the fissionable material in these devices has a shelf life of five years. It is believed that the Russians (the former Soviet Union) have not replaced any of this core material since the early 1980s. The suitcase device could possibly be used, however, to manufacture a radiological bomb.

A radiological bomb—or *dirty bomb*—is a conventional device built using military-grade explosives with nuclear material added. Detonation does not result in a nuclear explosion but spreads radioactive material throughout the target area. The size of the area affected would depend upon the amount of explosive and radiological materials used. This is a real threat, but the target area would most likely be a major population center rather than a utility facility.

According to the president of the Institute for Science and International Security, David Albright, there is a 10% to 40% chance that terrorists will conduct a successful attack with a dirty bomb before the year 2012.[6]

Chemical and biological weapons have been used by terrorists but, again, not against utility targets. These materials are relatively easy to obtain and result in widespread reaction among the target population, as was witnessed during the Sarin gas attack in Japan in 1995 and the anthrax attacks in the United States in 2001.

Summary

Terrorists and other extremists in the United States and around the world have targeted utilities and will continue to do so. But it is important to keep in perspective that the most likely threats to your facilities are from disgruntled employees and angry customers. If you are targeted by an extremist organization or a lone individual, it will most likely be a domestic terrorist using conventional weapons. Most of these attacks would utilize bombs or arson; other incidents might include sabotage or vandalism. A basic security program will help to protect against these threats. However, your utility should also be prepared to initiate the higher security levels outlined in the NERC's plan and the Homeland Security threat alert levels.

Notes

1. Seger, Karl A., *The Anti-Terrorism Handbook* (Novato, CA: Presidio Press, 1990)

2. Reeve, Simon, *The New Jackals* (Boston: Northeastern University Press, 1999), p. 154

3. Phelps, Timothy M., "Insight into Terror/Can't Stop Bombers, jailed Palestinian warns," *Newsday*, 05-17-1996, p. A-4

4. "Hackers Hit California Power Grid Computers," *Reuters News Service*, 06-09-2001, *http://ask.elibrary.com*

5. Miller, Judith, *et al.*, *Germs* (New York: Simon & Schuster, 2001), pp. 15–33

6. Nartker, Mike, "Radiological Weapons: 'Dirty Bomb' Attack is 40 Percent Probability," Expert Says, *Global Security Newswire*, 11-19-02, *http://nuclearno.com/text.asp?4431*

3

Conventional Threats

Workplace Violence Prevention Model

As illustrated in Table 1–1 in chapter 1, there are at least seven potential sources of workplace violence. Attacks by terrorists and other extremists are the least likely to occur (but have the greatest potential consequences).

Your utility cannot maintain medium or high threat levels (ES-Physical-Orange or ES-Physical-Red) for any extended period of time to address these extreme threats without serious operational and expense consequences. What's needed is a basic, day-to-day security program in place to protect your employees, customers, and assets from the *other* six sources of potential violence.

There are three categories of conventional threats:

- Insider threats
- Threats from outside the organization
- An insider operating in collusion with an outsider

The insider threat includes an enraged employee, ex-employee, or disgruntled employee. Outsider threats include angry customers, robbers, burglars, and vandals. Deranged individuals could be either an insider or an outsider.

Insider threats

The two primary insider threats are disgruntled and enraged employees. The disgruntled employee may not openly express his or her feelings and may be a greater threat than an enraged employee. The disgruntled employee quietly plans to get even for some perceived injustice and will either commit an act of sabotage or steal from the utility. In organizations where morale is low, there may be a number of disgruntled employees operating in collusion with one another. In this case, the value of the stolen assets can be considerable.

Enraged employees are easier to identify. They are both angry *and* vocal. They may be angry as a result of a perceived injustice, or angry with an individual, or both. The enraged employee may threaten others and may actually get into physical confrontations. Enraged employees have also been known to act out their emotions by sabotaging utility equipment or the property of the person who is the target of their anger.

In some cases a *disgruntled* employee evolves into an *enraged* employee. Workplace violence prevention professionals recognize some of the potential signs of this transition. They include:

- experiencing extreme personal and/or job-related stress
- minimal support system to deal with stress
- substance abuse
- frequent disputes with coworkers or supervisors
- violation of company policies
- harassment of coworkers
- threats of violence
- expectations of being fired or laid-off
- extremist opinions
- sabotage of equipment or processes

It is important to note that if a disgruntled or enraged employee, one who has been displaying signs of his or her discontent for some time, becomes quiet and withdrawn, it may be because this person has decided upon a plan of action. This is a critical time. The employee's behavior should be closely monitored.

Although he or she may no longer work for the utility, an enraged ex-employee is considered an insider threat. This individual knows your facilities and operations and may have a reason to "get even." In one case, failure to follow the security procedures that were in place resulted in two fatalities. An ex-employee who been terminated the preceding week returned to the utility on Monday morning and told the receptionist that he had some additional personal items to pick up. Despite the fact that the security procedures at the utility require that all visitors be escorted into the building and monitored, this employee was let through a locked door into the major office area without an escort and without having his story verified. The reason he returned on Monday was because he knew his former workgroup, including the supervisor, would hold its weekly meeting. He went directly to the meeting room and shot the supervisor and another supervisor who happened to be in the room. The shooting took place in front of seven other employees.

An employee who is experiencing domestic violence at home is in danger in the workplace, especially if the employee has left the abusive situation and is taking legal action against the abuser. There have been a number of cases where the abuser goes to the workplace to "get even." Employees in this situation should be urged to discuss the situation with either the human resources or security department. It may be necessary to move the employee into a temporary position and a secure location until the situation is resolved. Do not put another employee in the workspace vacated by the abused employee. The abuser may come to the workplace and shoot into the workstation or office without looking to see who is there.

Managing insider threats begins with the hiring process. Someone who has been a problem in prior workplaces is likely to be a problem in the future. Screening of applicants and background checks on people you intend to hire will help control the insider threat.[1]

Good supervision and fair disciplinary procedures are also needed. Supervisors should recognize potential signs of threatening behavior and the need to intervene before the threat escalates. When fair disciplinary procedures are in place, employees understand there are consequences for their actions, and most will be less likely to overreact if they are disciplined.

The Employee Assistance Program (EAP) is a valuable tool in addressing this threat. Again, the supervisor plays a critical role in making this tool effective. Supervisors should be able to recognize that an employee's performance or behavior has changed, and that the changes may be related to personal problems, substance abuse, or both. The supervisor must either discuss these concerns with the employee or report the individual to the human resource department or the EAP provider.

Basic security procedures also help address this threat. Access-control measures should be in place to protect the critical assets of the utility from disgruntled or enraged employees. Keys to offices and warehouses should have different levels of access. A master key will have limited distribution to open most offices and storage areas; other keys will open selected offices or other areas. Keys should be distributed like sensitive information, on a need-to-know (or to gain access to) basis.

In addition to access control, there should be assets control. If you leave your chainsaws lying around in uncontrolled areas of the warehouse, some of them will disappear. If tools are not accounted for by vehicle, employee, or crew, you will be purchasing new tools on an ongoing basis. Some companies lose thousands of laptop computers each year because employees are not held accountable for their loss. Assets control is achieved by securing assets that would be attractive to employees, by controlling their use and distribution, and by holding employees accountable when items are lost.

Outsider threats

The number one outsider threat to distribution utilities is enraged customers. They are a threat to employees and to utility property. Unfortunately, there will always be angry customers. Utility employees are able to defuse many of these threats, but, in some cases, law enforcement officers may be needed. Inside employees are protected using access controls, closed circuit surveillance systems, and other office or building security measures. Outside employees are at a greater risk, since it is not possible to apply these measures outside of the office.

A revenue protection agent from an electric utility was taking photographs at a tampered meter when the customer leaned over a balcony and emptied a .22 caliber rifle in his direction. Fortunately, the utility employee was not hit, and the shooter received six months in jail. In another theft-of-service case, a lineman was sent to a rural area to disconnect the customer at the pole. After the lineman climbed the pole, the customer walked up with a shotgun in hand. The utility employee was able to maintain control of the situation and eventually talked the customer into letting him climb down and leave.

Part of the threat in the field is that you don't usually know who the customer is or what the degree of the threat may be. An African-American meter reader found a service where the meter had been removed and wires placed into the socket. The customer came out of the house with a gun and told the meter reader to get off his property. When the police arrived, the meter was still out of the socket, and the customer was arrested. When they checked his criminal record, they found the man had been in prison twice for two separate homicides and was the leader of a local white supremacist group.

Some customers experience spontaneous rage—they suddenly become angry over a service or billing problem. Customer service representatives can usually deal with these problems and defuse the situation. But anger does not dissipate as quickly as it begins. When a person becomes angry, certain hormones are released into the body, and the nervous system becomes highly activated. That is why giving an angry person the opportunity to vent is important. It helps to bring this level of activation back to normal, or close to normal, so that you can deal rationally with the customer.

Other customers become angry over a period of time. They are more dangerous than those experiencing spontaneous anger because they have time to plan before you know the situation exists. A person experiencing spontaneous anger doesn't have time to build a bomb or plan arson. The customer who experiences gradual anger and rage does. Not all of these customers become physically violent. A physician who became angry with his utility over several issues decided to get even by parking his car on top of the meter cover so that it could not be read. He became angrier when utility employees arrived with a sheriff's deputy and a tow truck and removed the car from the meter.

Just as utilities will always suffer with enraged customers, they will also become involved with civil disputes. The most common of these is right-of-way disagreements. In one case, a farmer, upset over a right-of-way dispute with a transmission utility, threatened to shoot the next utility employee he saw on his property. That employee happened to be a staking engineer who was not told of the farmer's threat. The farmer shot and instantly killed the staking engineer.

Civil disputes usually result in highly emotional situations. It surprises many people when they learn that most shootings in courthouses take place during civil rather than criminal trials. This is due in part to the security at criminal trials, but it is also the result of the highly emotionally charged atmosphere that is created during a civil trial. People who become emotional may also become irrational—and dangerous.

Utilities should learn as much as they can about customers and others who initiate a civil action against them or defendants against whom the utility initiates the lawsuit or other action. In addition, every threat against the utility or its employees must be recorded and responded to. A utility that had recently built a new headquarters building with a glass facade was told by an irate customer that he had "plenty of rocks" and that he would be using them the following weekend. The utility reported the threat to the police department but was told there was nothing the department could do until an incident occurred. Since the police refused to take the lead, a utility manager telephoned the customer. The manager discussed the situation with the customer in a professional manner and told him threats against the utility or its employees were not acceptable and were reported to the police department. The manager encouraged the customer to telephone him personally if he had any future problems or concerns.

Utilities are concerned about the possibility of robbery in the office and in the field. If your utility is still collecting cash from customers in the field, consider changing your policy. If you have to collect payments in the field, accept only checks or money orders.

Let's put this into perspective. Most customers who do not pay their bills until the disconnect notice or a field services employee arrives are not your best customers. Yet many utilities still believe they have to provide a service to these people by accepting payment at the service location, including cash payments. This creates two potential problems. The first is the possibility of the field employee being robbed. Many of these employees are carrying more money at the end of the day than most robbers get during a convenience store robbery, and they are alone and vulnerable. The other potential problem is that the customer can claim to have paid the bill to the field employee but was not given a receipt, when, in fact, the customer did not pay the bill. If your utility collects payments in the field, then you are putting service to your least valuable customers above the safety of valued employees.

Robberies at utility offices are not uncommon. Any office where cash is handled should have surveillance cameras and a notification at all entrances stating, "This building is under video surveillance 24-hours a day." Additional procedures to help reduce the probability of a robbery at your offices are discussed in a later chapter.

Remember the third category of conventional threats—an insider operating in collusion with an outsider? A utility employee in a large city helped his friends plan a robbery. He knew when the most cash would be in the office and when it would be most vulnerable. The robber's net was in the tens of thousands of dollars. (The employee was caught when he became a "conspicuous consumer," spending his share of the money on a new car and other items that were beyond his income.)

Burglars also target utilities, and sometimes the utilities encourage them. Do you leave the keys in your vehicles if they are parked inside a fence at night? At one utility, burglars climbed over the fence and stole a pickup truck, which they rammed though the fenced gates to escape. The utility bought better locks for the gates but continued to leave the keys in the vehicles. The next time the burglars visited, they used a line truck to ram the gates, then stole another pickup.

At another utility, burglars stole three pickup trucks on three different occasions. The utility still leaves keys in the vehicles overnight. The question must be asked, who is the dummy here?

Offices are targeted by burglars for cash (including night deposits) and valuables, including computers and other office equipment. Good physical security and intrusion detection systems help to address this threat.

Burglars target warehouses and other storage areas for tools, copper, and other valuables. Physical security, security lighting, and intrusion detection systems reduce the probability of your utility being targeted.

Another outsider threat to your utility is extortion. In rare instances, a criminal may threaten to disrupt your services if certain demands are not met. In today's world, however, most extortionists threaten your information technology (IT) systems. If your utility offers Internet Service Provider (ISP) services, then the extortionist will target the credit card numbers stored in the system.

Threats of collusion—insiders and outsiders

The disgruntled employee is most likely to scheme with outsiders to vandalize or sabotage your utility. This employee knows where you are most vulnerable and knows what points in the system are most critical. These incidents can be devastating.

Several employees at an electric utility were disgruntled not just with the utility but with the government system in general. As members of an extremist group, they plotted with several outsiders to shut down the hydro-generation capability at one of the utility's dams while also opening the spillways and flooding downstream. They were arrested on other charges before they were able to carry out their plot.

For other insiders, the motive is profit; the employee who provides the information needed for others to rob his or her employer of a considerable amount of money isn't disgruntled—just greedy. In another case, employees or contract cleaning crews take note when a utility receives a large shipment of copper conductor. The following weekend, the warehouse is burglarized. Unless the employees or the cleaning crews become conspicuous consumers, their connection to the burglary may never be discovered.

Deranged individual

The deranged or emotionally disturbed individual may present an insider or an outsider threat. Many conditions can lead to these behaviors, including personality disorders and paranoid schizophrenia.

Personality disorders include inadequate personality and antisocial personalities. Individuals with an inadequate personality disorder have trouble controlling emotions. They usually know right from wrong but are able to rationalize the wrong choices they decide to make. They are not necessarily violent but may become involved in criminal behavior.

The *antisocial personality* is more likely to commit crimes. This person fails to recognize the difference between right and wrong and may have little control over emotions or behaviors.

Paranoid schizophrenia is a psychiatric disorder, which can usually be controlled through medication. Problems occur when the patient fails to take the medication as prescribed. This person typically hears voices that direct behaviors and is compelled to follow the instructions given by these voices.

When confronted by a mentally disturbed or deranged individual, your primary objective should be to get away as quickly as possible. Remain calm and in control, but disengage as soon as you can. Don't believe anything that person tells you. Try to notify the police so they can intervene.

Threats to Utilities

Utilities are constantly confronted by threats from natural disasters and deliberate attacks. As a result, they can usually respond quickly and appropriately when an incident or disaster occurs. But the proactive function of a security program is to prevent some of these incidents. Individuals or groups responsible for deliberate attacks against utility systems include vandals, criminals, ecological extremists, bored hunters, and terrorists.

As utility systems become more reliant on automation, they also become more vulnerable. Consider the following potential threats to power control systems as an example:[2]

- Access via the main information system
- Access via other utilities or power pools
- Links to supporting vendors
- Remote administration ports
- Potential to crash the system
- Corrupt databases
- Issue false commands
- Manipulate the flow of data to the center

Threats to substations include:[3]

- Dialing into an unprotected port on a digital breaker and changing the tolerance
- Accessing a remote terminal unit (RTU) and issuing commands or reporting spurious data
- Bomb, arson, or intentional shorting of substation components

Electric utilities are not the only systems at risk. There has been a concern since September 11, 2001, regarding threats to water systems. The major threat has been the possible damage of water supplies by destroying a dam or other water containment structure. Other concerns include contamination of the water supply, loss of pressure resulting from the destruction of water mains, destruction of pumps, and targeting chlorine at treatment facilities.

There are more than 900,000 oil and gas wells in the United States. It is impossible to protect all of them. In addition, there are 2000 petroleum storage terminals, 161 petroleum refineries, 726 gas-processing plants, and 410 underground natural gas facilities. Each of these represents a potential target for a deranged individual or an extremist group.

Summary

Utilities are confronted by the security threats that all businesses face. These include disgruntled and angry employees and customers, as well as armed robbers and burglars. However, because of the unique nature of utilities, utilities are also targeted by vandals, extremists, and others wanting to attack an infrastructure.

Security is important at all businesses, but, for utilities, adequate security, including plans for increased security levels, is an absolute necessity for protecting the system, customers, employees, and assets.

Notes

1. Seger, Karl A., "Violence in the Workplace: An Assessment of the Problem Based on Responses from 32 Large Corporations," *Security Journal*, vol. 4, no. 3, July 1993, pp. 139–149

2. "The President's National Security Telecommunications Advisory Committee Assurance Task Force Electric Power Risk Assessment," March 1977, *http://www.securitymanagement.com*

3. Oman, Paul, *et al.*, "Concerns about intrusions into remotely accessible substation controllers and SCADA systems," Schweitzer Engineering Laboratories, 2000, *http://www.selinc.com*

SECTION 2 Managing Threats

4

The Risk Management Process

Risk Analysis

Several risk analysis equations are available. One of these was developed at the Sandia National Laboratories and is used in utility risk analysis. The Sandia General Risk Equation should be computed for each potential undesired event in this way:[1]

$$R = P_A \cdot (1 - P_I) \cdot C$$

where:

R = risk associated with adversary attack
P_A = likelihood of attack
P_I = probability of interruption
C = consequence of the loss from the attack

The Sandia Risk Equation is based on extensive research and real-world applications by the DOE and other major organizations. Application of the model, however, may prove daunting for many utilities. The formula should be computed for *each* potential undesired event at each facility, and it requires the team conducting the risk analysis to estimate the probabilities of the likelihood of attack, the effectiveness of the security system against the attack, and the probability that the attack would be successful.

There are two other assumptions upon which this approach is based. The first is that you are protecting material that is of national security interest. The Physical Protection System (PPS), from which the Sandia Risk Equation is derived, was designed to protect DOE sites where special nuclear material or classified information is located. Other than nuclear generating plants, few utility facilities contain material that is of national security interest. Second, the security systems at PPS sites include an on-site armed response force that is trained to respond to and repel attempted intrusions or assaults. Your utility may have uniformed security personnel in the cashier lobby, but seldom are guards, armed or otherwise, stationed at other utility facilities, including electric substations, water and wastewater treatment plants, or gas storage or compressor stations.

The Sandia PPS is excellent for high-threat targets when applied by trained security professionals. However, there is a simpler risk equation that can be used for most utility risk analyses:

$$Risk = Threat + Criticality/Vulnerability \quad (R = T + C/V)$$

where:

Threat	is the potential threat to a specific location.
Criticality	is the potential consequence of a successful attack or other untoward event including natural disasters.
Vulnerability	is the effort required to disrupt operations at this facility.

The $R = T + C/V$ approach requires only one risk equation for each facility, although it will also be performed for specific locations within a facility, such as the control room. It does not require a separate formula for each potential undesired event at each facility. It also avoids the need for an assessment team to provide numerical estimates (low = 0.1, medium = 0.5, and high = 0.9) for the probabilities included in the Sandia Risk Equation. Again, the Sandia approach is excellent for high-threat facilities but not as appropriate for many utilities that are at less risk and where risk analysis is conducted by operational personnel rather than security professionals.

As risk increases, the threat condition level at the utility should increase correspondingly. The NERC and the Homeland Defense systems have five. The number of risk levels your utility must be prepared to address depends upon the threat alert system you adopt.

In your risk analysis process, a number of different categories of risk should be considered. These include risks to the following:

- ***Property*** reduction in value or loss
- ***Liability*** responsibility for loss by others
- ***Personnel*** safety, disability, death, or reduced efficiency
- ***Physical*** destruction or damage
- ***Social*** impact on an individual or group
- ***Market*** changes in price, value, or competition

The risk analysis should be conducted by a team including operations specialists and security personnel. Smaller utilities that do not have security personnel on staff or the ability to contract for these services may be able to work with crime prevention officers from a local law enforcement agency. For utilities with multiple locations, the first task of the team is to determine which of these locations are critical to utility operations. If an electric utility has 20 substations but only 4 of these are connected to the transmission grid, do they all have the same level of criticality? Obviously the 4 substations connected to the gird are probably more critical than the other 16. The team should completely map the utility system, determine which points are the most critical, and then focus on conducting a risk analysis of each of these points. An example of the criteria used to determine criticality at specific locations is discussed later in this chapter.

The team may determine that an event at some of the locations on the system may result in a temporary disruption of service but that these points are not critical to overall system operation. Their risk analysis report should identify these locations and state why the team did not consider them critical. Risk analyses will not be conducted at these locations.

You may consider initiating security standards at noncritical locations. These could include specific deadbolt locks on doors, screening or bars on windows, and minimal fencing around structures. These standards would be applied to all similar noncritical points on the system.

Threat analysis

Threat analysis is not limited to extremist threats or even security threats. As you conduct the analysis, also consider equipment failures, natural disasters, and other potential events. For our purposes, we will focus on intentional acts.

It may not be necessary to conduct a threat analysis for each facility. Distributor utilities may only need to conduct an analysis for their primary service area. Companies with multiple distribution service areas and transmission systems, and others with regional or national interests, should conduct a threat analysis for each geographic location.

Begin your threat assessment by examining previous incidents and threats to your utility. If you have been targeted for robbery or burglary in the past, then you have valuable assets that may result in your being targeted again. If disgruntled customers have vandalized your property in the past, expect them to return. Your utility should keep a record of all threats and criminal events targeting the utility. This record can be established using a database management program. There are several commercial software programs available for this purpose.

In chapter 3, we discussed the workplace violence prevention model. This is the second step in your threat assessment. Review with management each of the seven components of the model and the possible threat from each. Consider entering the components of the model and the potential threat discussed for each category into a table in your word processing software and including this in the risk analysis report (Table 4–1).

Table 4–1 Workplace Violence Assessment

Categories of Workplace Violence	Potential for Incident at this Facility
Enraged (Ex) Employee	
Domestic Violence	
Angry Customer	
Civil/Legal Dispute	
Robbery or Other Crime	
Deranged Individual	
Terrorist or Hate Crime *(include specific interest group threats)*	

Several steps of your threat assessment will be completed on the Internet:

- an assessment of the demographic characteristics in your service area
- crime statistics for the area
- identification of extremist groups that may present a threat to your utility

What do you know about the community you serve? How does it compare to other communities in the state and the nation? The threat assessment includes an examination of the census data for your service area. Some of the data you may wish to examine includes:

- change in population from previous census (are we growing or declining?)
- household ownership rate and medium household income as compared to state statistics
- persons below poverty as compared to state
- personal income as compared to state

A number of other statistics may help you to better understand the demographics of your service area. These are found at *www.fedstats.gov*.

Is crime in your service area increasing or decreasing? If there is a change, is it related to violent crime or property crimes? Utilities are no different from any other organizations in your area. If burglaries are increasing at other businesses, then you are at greater risk than in the past. If violent crime is increasing in your jurisdiction, then it is an increasing threat to your employees and customers.

The next step of your threat assessment is to return to the Internet to examine the crime information for the cities and counties in your service area. Local crime information is located at the Bureau of Justice Statistics web site. To find the data for your service area:

1. Begin at *www.ojp.usdoj.gov/bjs/*
2. You should now be at the home page for the Bureau of Justice Statistics. Look for the section to the right on "Data for Analysis," and click "Data Online."
3. You are now at the "Crime and Justice Online" page. Click "Crime trends from the FBI's Uniformed Crime Reports."
4. On the Crime Trends page, click "By local reporting agency."
5. On the Local Level Crime Trends page, click "Single agency trends."
6. Choose state and population of the jurisdiction. Click "Next." You can identify the population using the census data previously downloaded from the fedstats web site.
7. Choose crimes variable groups (suggest you choose all groups) and agency. Click "Table."
8. Print crime statistics. Make sure to use the Landscape printer option. You may have to decrease the font size to print all of the variables.

(*Please note:* This sequence of data entries works as this is written. The Internet changes daily, so if you have problems, modify your entries. However, the criminal justice data is somewhere on the Bureau of Justice Statistics web site!)

As a final exercise on the Internet, identify extremist groups that may be located in your service area. Visit three Internet sites to help determine if and where these groups may be located. Begin with the U.S. Department of Justice site at *www.usdoj.gov*. Click "search" and enter the following searches:

- extremist groups and (name of your state)
- terrorism and (name of your state)
- crime and (the name of your state)

Next, go to any major search engine such as *www.google.com*. Enter the same search terms. Open and print those documents that seem relevant to your threat assessment.

Finally, go to the web site of the Southern Poverty Law Center (SPLC). This organization tracks anti-government militia groups and hate groups. You will find a listing for each category on the Intelligence Project pages of the SPLC site. Groups are identified by name and location. You may want to go back to the google.com site to research any groups the SPLC lists in your service area. The SPLC web site is at *www.splcenter.org*.

Please note that many of the groups listed on the SPLC Intelligence Project may have extremist opinions but are not engaged in criminal activity. In most civilized countries, expressing extremist views is not a crime, and these people and groups are exercising their rights. Occasionally, however, someone is influenced by one of these groups and commits appalling crimes. Timothy McVeigh was influenced by at least one extremist group, as was his co-conspirator, Terry Nichols. Benjamin Smith, a member of an extremist group headquartered in Illinois, went on a rampage, shooting six Orthodox Jews emerging from religious services and an African-American minister at his church. He then fired four shots into a Korean church. Two people died and nine were injured before Smith, surrounded by police, committed suicide.

What weapons or tactics would an adversary use to target your facilities? In some areas, there is more gun ownership than in others, partly due to local laws and partly due to the local culture. If there is mining in your area, there is probably a greater access to explosives and people who know how to use them. To assess this part of the threat, the threat committee should discuss each of the variables in Table 4–2 with local law enforcement to determine which represents a potential threat to your utility.

In chapter 2, we discussed the criteria that terrorists and other extremists use to select potential targets. How do your facilities match these criteria? Table 4–3 lists the criteria in the left-hand column. Make notes in the right-hand column that describe how the facility does or does not meet these criteria. You need to do this for each critical facility.

Table 4–2 Potential Weapons and Tactics

WEAPON/TACTIC	POTENTIAL AT THIS FACILITY
TRADITIONAL TACTICS	
Bombing or Bomb Threats	
Arson	
Murder or Assassination	
Armed Attack	
Hostage-taking	
Kidnapping	
Sabotage or Vandalism	
Other Threats	
WEAPONS OF MASS DESTRUCTION (MEGATERRORISM)	
Chemical Agents	
Biological Agents	
Nuclear Devices	

Table 4–3 Extremist Target Selection Criteria

TARGET SELECTION CRITERIA	APPLY TO THIS FACILITY
Criticality of the Target	
Accessibility to the Target	
Media Value of the Target	
Recoverability of the Target	
Vulnerability of the Target	
Potential Backlash to the Terrorist Group	
Risk to the Tactical Cell	

The process just discussed should help evaluate threats to specific locations on your system. If all of your facilities are in one service area, the process may only need to be completed for that area. If you have a number of facilities across the country or region, a series of threat assessments needs to be conducted. If your company has international interests, then country and regional threat assessments are required.

Some companies develop a generic threat statement to complete their threat assessment. An example of a threat statement for an NERC-licensed facility may include the following language:[2]

> *The threat is considered to include a determined, violent external assault, attack by stealth, or deceptive actions by persons with the following attributes:*
>
> - *Well-trained and dedicated individuals*
> - *Inside assistance from knowledgeable employees, former employees, or vendors*

- *Suitable weapons to include automatic and long-range weapons*
- *Explosive and incendiary devices*
- *Vehicles for transporting personnel to and from the site*
- *The ability to operate as two or more teams*

This statement provides a basis for determining what countermeasures are needed to manage these threats. However, the statement is generic and does not address other threats such as a disgruntled employee who goes berserk. If an employee injures or kills other employees or customers, and this threat was not in your "company threat statement," surviving relatives may sue you, and this omission will probably be pointed out in court. A generic threat statement can be used as a working document, but don't make it part of the organization's threat assessment.

Determining criticality

Determining criticality should be the result of your consideration of the criteria listed in this chapter. It may also be determined by regulations governing your operations. There are specific criteria (for instance) for utilities regulated by CFR 49 Part 192 that require gas pipelines to identify their DOT class. Different classes of facilities have different requirements under the regulation. Classes range from class 1, "an offshore area; or any class location unit that has 10 or fewer buildings intended for human occupancy," to a class 4 location, "any class location unit where buildings with four or more stories above ground are prevalent."

If you are located within the United States at a facility containing more than 10,000 pounds of hazardous substance, then you are familiar with the Environmental Protection Agency's (EPA) Risk Management Program and the Occupational Safety and Health Association's (OSHA) Process Safety Management (PSM) program. Facilities that are regulated by DOT, EPA, OSHA, and others will most likely be on your critical facilities list.

Large utilities have hundreds of potential target locations, and it is impossible to conduct a risk assessment and countermeasure for each of these. There must be a systematic approach to determining which facilities are truly critical and which are not. There are a number of different criteria that can be evaluated when determining criticality. These include:

- Impact on international and national infrastructure
- Impact on regional and local infrastructure
- Impact on community population
- Environmental impact
- Business/operations disruption
- Impact to facility population
- Presence of hazardous materials
- Consequence management considerations

Considering possible consequences of an event at a specific location forces us to keep the criticality process simple by classifying each facility into one of three basic categories:

- ***Highest*** essential to operations (Point value = 3)
- ***Medium*** can operate without for a short duration (Point value = 2)
- ***Low*** can operate without for extended period of time (Point value = 1)

Let's apply this approach to the eight criteria for determining criticality to derive a numerical criticality value (Table 4–4).

Consider the impact on international, national, regional, or local infrastructure by reviewing the cascading effect of a failure to an infrastructure component. If electric service is disrupted to a local community for an extended period of time, what are the cascading effects? Water treatment and wastewater treatment plants cannot operate. Hospitals and other healthcare facility have back-up systems, but for how long can they operate? Do other critical facilities have back-up systems that can operate for extended periods of time without electric service? In many cases, critical infrastructure components are considered from an individual perspective. But in reality, when one component of a critical infrastructure is

Table 4–4 Criticality Assessment

CRITICALITY CRITERIA	CRITICALITY POINT VALUES		
	ESSENTIAL TO OPERATIONS	*CAN OPERATE W/O FOR SHORT DURATION*	*CAN OPERATE W/O FOR EXTENDED PERIOD OF TIME*
Impact on International and National Infrastructure	3	2	1
Impact on Local Community Population	3	2	1
Environmental Impact	3	2	1
Business/Operations Impact	3	2	1
Impact to Facility	3	2	1
Population	3	2	1
Presence of Hazardous Materials	3	2	1
Consequence management considerations	3	2	1
Total for Each Criterion	Sum of Column	Sum of Column	Sum of Column
	Total Criticality Assessment		Total Score

affected, it has a cascading effect on other critical infrastructure components. Potential cascading effects should be considered and plotted when conducting a criticality assessment.[3]

All traffic lights are out in your service area because the electric service has been disrupted. What is the impact on the local population? In Loudon, Tennessee, the impact is minimal because there are only a few traffic lights, and local police would readily respond to the situation with traffic control officers. In a major city, it is a different situation. There is a lot more traffic, and traffic lights are critical for control. Police may be responding to other security needs such as crowd control, and there will be fewer officers, per population, available for traffic control. The situation could be critical.

Your primary obligation is to provide utility service in your area. If a critical point in your system is attacked and service is disrupted, what is the impact on your business and its operations? How soon will you be able to restore service? What revenues are lost as a result of the attack and disruption, and how will you recover them?

If a bomb explodes at one of your facilities, what is the potential loss of life? If a hazardous material is released, how many people will be affected? A key consideration in the criticality assessment is the potential effect of an incident at a specific location on the people who work at that facility.

If a bomb explodes at an electric substation, it may disrupt service in an isolated area for a limited time. But if the same bomb explodes at the chlorine storage areas of a water treatment plant or at a natural or propane gas facility, the damage could be much greater. If a specific location has hazardous materials, its potential for being classified is increased. As discussed, if more than 10,000 pound of hazardous materials is at the site, then it is a *critical* site according to U.S. regulations.

For any incident that occurs at any site, what would it take to respond? This is the consequence management consideration. The greater the potential consequence of an event at a facility, the greater its criticality. *Criticality* is sometimes referred to as the severity of the consequence of an event.

Applying these concepts to Table 4–4, the risk assessment team should discuss each of the eight criteria in the table and determine point values. The result will be a total number of points on the criticality scale ranging from 8 to 24. If a facility rates 8 points, it is not critical. If it rates between 9 and 16 points, it is probably not critical, but the team should agree on this classification. If the facility rates 17 to 24 points, it is probably critical and you should conduct a risk assessment. If a facility rates 24 points, it is *definitely* critical.

Your team has conducted a criticality assessment for each facility; now you should identify the specific critical components at each facility. A number of components at each facility are not critical to its operation. Others are critical. Whether we are evaluating an electric substation or a gas compressor station, the operations building will be a critical component. For each of the facilities considered critical, conduct an on-site criticality assessment to determine the critical components at that specific facility. Simplify the process by using our criticality approach.

Now focus on the critical components at the facility as we begin the vulnerability assessment (VA).

VA

There is a misunderstanding in some segments of the utility industry in regard to VAs. In some cases, the total focus of the risk assessment is on determining vulnerability, whereas, according to our formula, $R = T + C/V$, vulnerability is one part of the total equation. True risk can never be identified if the direct threat and the criticality to each site are not determined first.

In this section, we will discuss different steps to the VA using several tools. Remember, however, that a VA will only be conducted, in most cases, on facilities that have been determined to be critical. Further, the most effective VAs are conducted on facilities where the local threat has been identified using the threat assessment approach.

A useful starting point for the VA is the Facility Vulnerability Determining System (FVDS) (Appendix B). The FVDS is based on an assessment tool developed for the DOD in the 1970s and has been updated for use in determining vulnerability to facilities in today's world. While the FVDS generates a point value ranging from 0 to 100, do not develop a "points mentality" when using the instrument. A high score doesn't mean an attack is imminent. At the same time, a low score is not an indication that your facility will not be targeted. The primary value of the FVDS is to evaluate the facility vulnerability on each of the 11 criteria and to note why specific point values are allocated on each of these. In many cases, the vulnerability on a certain criteria cannot be changed (the geographic region where the facility is located, for instance). Others *can* be addressed, such as status of training and security awareness. The 11 criteria included in the FVDS are:[4]

1. Facility characteristics and sensitivity
2. Status of training and security awareness briefings
3. Available emergency communications
4. Availability of law enforcement resources
5. Time and distance from other facilities or organizations with mutual response capability
6. Time and distance from urban areas
7. Geographic region
8. Population density of the facility
9. Proximity to foreign borders
10. Access to the facility
11. Terrain around the facility

During the threat assessment, you identified potential weapons and tactics that could be used against your facilities. Next, you determined which facilities are critical. In this part of the VA, you should assess which of these tactics could be used against each of the critical facilities and the potential consequences of their use. A well-placed bullet could cause extensive problems at an electric substation, but what components at the substation are most vulnerable to this event? What are the security weaknesses that would allow this event to occur? Are there countermeasures that could reduce this vulnerability?

At each critical facility, identify the most vulnerable points and the weaknesses making them vulnerable. There is a criticality/vulnerability interaction consideration that is important to this process. A tool shed may be located outside of the perimeter fence and be vulnerable to burglary. But if there is nothing of value in the shed, do we care? If the tool shed is not critical to the operation of the facility, its vulnerability is of less interest. When conducting the vulnerability assessment, first determine which points or processes at each facility are critical, using the same criteria used to identify which facilities are critical, and then assess the vulnerability of each of these points or processes.

Think like the adversary. If you were going to target the facility, what specifically would you target, and what weapons and tactics would you use?

The PPS developed at Sandia Laboratories uses fault tree analysis in the VA. A separate fault tree is developed for each potential untoward event, and each of the steps leading to the event is identified. A complete fault tree analysis for a water treatment plant, electric generating station, liquefied natural gas (LNG) storage site, or other major utility facility could be an extensive activity, but the results should help to identify each threat and the multiple vulnerabilities leading to the threat. Fault tree analysis may not be needed at smaller facilities where the vulnerabilities are obvious. An excellent resource on PPS is a book by Mary Lynn Garcia.[5]

There are two other tools developed for the PPS that should be used during the VA: the estimate of adversary sequence interruption (EASI) and the adversary sequence diagram (ASD).

The ASD is the most likely route an adversary would take to the critical target at a facility. The ASD does not have to be to scale, but it should show each of the potential pathways to the target (Fig. 4–1).

The primary purpose of the EASI is to evaluate the security measures in place, to determine the probability of an attack being successful, the probability of detection, and the mean response time to attack. A computer model for this analysis is available on the DOE web site. For our purposes, let's simplify this process.

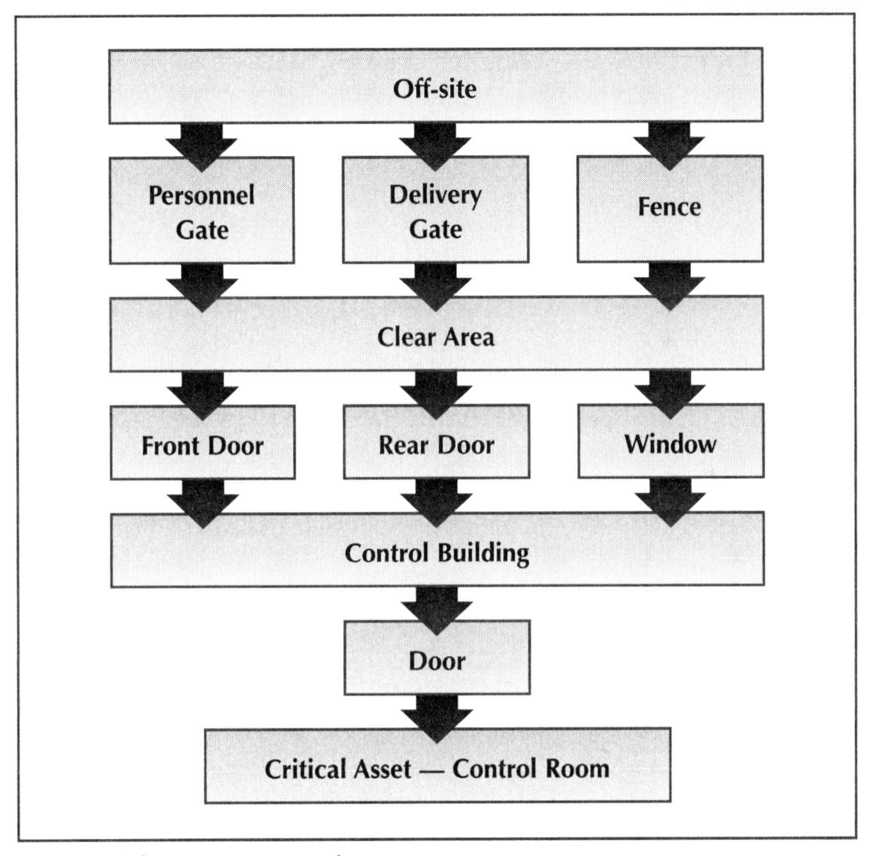

Fig. 4–1 Adversary sequence diagram

Assume the critical target is an operations building located at a critical facility. Answering five questions will help you identify the building's vulnerabilities.

1. At what point would adversaries come through the perimeter fence, and how? Would they cut through the fence and run to the target, or would they crash through the fence gate with a vehicle and drive to the building?
2. What is the probability they would be detected before reaching the control building?
3. If they were detected, how much time would be required for law enforcement or a guard force to intercept the adversary?

4. Assume they make it to the building and attempt to enter; what is the probability they would be detected, and how much time would the response force require?
5. If they got into the building, how much time is required to reach the control room? Would they be detected? What is the time required for the response force?

A key to the model is the time required for the adversary to complete the event. How long does it take to cut through the fence and run to the building? How long does it take to gain entry, and how long does it take to gain access to the control room? The EASI computer model simulates the intrusion repeatedly, and computes the mean and standard deviations of the time intervals. The model also computes the *probability of guard communication* and the *probability of interruption*. Let's consider the path of the attack and simulate it at least once using a simplified EASI model. These data should be collected at the critical site and not estimated using site plots or maps:

1. Begin at the fence. If you were the adversary, how would you enter, and how long would it take? You don't have to actually cut the fence or crash a vehicle through the gate. Just estimate the time required.
2. Would you be detected by an intrusion alarm system, security surveillance camera, or an alert employee?
3. If you are detected, how much time is required for law enforcement or security to intercept you?
4. How long does it take for you to move from the fence to the building?
5. How much time is required to enter the building? Is it a secure structure that is kept locked or a structure with numerous windows and doors that are *not* locked?
6. Once in the building, what is the time required to reach the control room?

You now have the time required for an adversary to reach the critical point (*i.e.*, the control room), you have assessed the probability of detection at each point, and you know the response time required for law enforcement or security to intervene once the intruder is detected. The important questions—will law enforcement or security arrive before the intruder reaches the critical point, and what countermeasures can be initiated to delay the adversary? Countermeasures may be as simple as locking the doors at all times or may require the installation of intrusion detection systems at critical points. As stated earlier in this chapter, one of the assumptions related to the PPS is that you have an armed guard force on site. But in most situations, you can rely on local law enforcement to respond to an event. In many cases, the adversary will probably reach the target before the response force can intervene. Delaying access to the critical point, when possible, is your best defense (Table 4–5).

Table 4–5 Estimate of Adversary Sequence Interruption
(modified from the DOE model)

Facility	Electric transmission substation			
Critical Asset	Control room			
Task	**Description**	**Probability of Detection**	**Time Required (Seconds)**	**Police/Guard Response Time**
1	Cut fence	0	30	
2	Run to building	0.5	90	
3	Break through door	0.5	90	
4	Run to control room	0.9	20	
5	Sabotage controls	0.9	120	
	Summary	Not likely until adversary reaches the building	350 seconds = 5.83 minutes	270 seconds = 4.5 minutes
	Conclusion	Adversary would be in the process of sabotaging the control room before police or guards respond. Recommend additional intrusion detection system to detect the intrusion earlier and an additional delaying strategy such as an additional fence around the control center.		

Risk reduction

Once you identify threats, criticalities, and vulnerabilities, the next step is to identify methods for reducing these risks. If you are in charge of your utility's security (or from another department but assigned to a security or risk reduction committee), remember that your job is to assess risks and develop risk-reduction strategies. It is then management's prerogative to decide how much risk is acceptable and to implement those recommendations they deem appropriate.

There are several considerations that determine the need and utility of a security risk-reduction measure:

- The security measure is required by law or regulation.
- The cost of the security measure is small, but its benefit is material, *e.g.*, instituting a procedure to secure a door providing access to a critical area.
- The cost of the security measure will be more than offset by the reduction in future losses. You can use the Annual Loss Expectancy (ALE) approach to determine if the security measure has a positive return on investment.
- The security measure addresses an event that has a low probability of occurrence but a high consequence if it does occur. In this case, you would use the Single Loss Occurrence (SLO) evaluation in your recommendations to management.

Begin your risk-reduction process by developing a physical protection and vulnerability matrix. The matrix has four components:

- Detection features
- Delay features
- Response features
- Mitigation/safety features

Assume that we are protecting a control building at one of your critical facilities. *Detection features* may include surveillance cameras near the fence, personnel working in the area, or neighbors who live or work in the area. *Delay features* include the fence around the facility, the distance from the fence to the building, and the building itself. *Response features* include the response time for law enforcement or security, or for employees working in the area. *Mitigation and safety features* include your recommendations for risk reduction. These may include locking the control room at all times or the installation of an intrusion detection or panic alarm system. An example of a physical protection and vulnerability matrix is shown in Table 4–6. You should complete a matrix for each critical location at each critical facility.

Table 4–6 Physical Protection and Vulnerability Matrix

Detection Features	Delay Features	Response Features	Mitigation/Safety Features
Security officer at entrance	Clear areas both sides of fence	Unarmed guard on site with radio	Additional lighting near fence
Surveillance cameras on fence	Perimeter fence *(6-foot chain link)*	Law enforcement response *(5.5 minutes)*	Roving guard
Personnel working in the area	Standard locks on all doors		Greater employee awareness

How cost-effective are your security recommendations? If your utility experienced a series of burglaries in the past year that resulted in $500,000 in lost assets, should you spend $1 million on security systems to reduce those losses? What about spending $10,000 if that expenditure will reduce your losses by an estimated 5%?

In the first case, spending $1 million to save $500,000, the utility loses $500,000 if they follow your recommendations. The utility is better off letting the burglars in. In the second case, spending $10,000 to reduce the loss to $475,000, you are still not meeting your objective to significantly reduce the loss using a cost-effective approach. There are two realities to most security programs. First, you will never reduce your risk or loss to zero no matter how much you spend on security. Second, if you don't have an effective security program, your losses will potentially be enormous. The challenge is to find the balance between the cost of security and risk and loss control.

Annual loss expectancy

The ALE of a risk is the product of its rate of occurrence, expressed as occurrences per year, and the loss resulting from a single occurrence.[6] For example, last year 1000 customers failed to pay utility bills after service was disconnected. If the average loss was $500, then the ALE is $500,000. To reduce this loss, the utility implemented a procedure requiring all new customers to show government-issued photo identification, either in the office or at the service location, before service was initiated. In addition, the utility installed a credit reporting system to verify credit and identify each new customer. The initial cost of the system was $100,000. The cost of maintaining the system for the next four years will be $30,000 per year. Is this a cost-effective procedure?

1. Annual loss is $500,000. Over the next five years, the loss could total $2,500,000.
2. Initial cost of the new system is $100,000, and the annual maintenance cost over the next four years is $30,000 per year ($120,000). Total investment is $220,000.
3. It is expected the new systems will reduce the ALE by at least 60%, or a total of $1,500,000. Losses over this period are expected to be $1,000,000.
4. You are recommending spending $220,000 to save $1,500,000. Is this a cost-effective risk-reduction program?

Figure 4–2 is an example of an ALE worksheet.

Loss Analysis Target ____________________

Annual Loss Expectance

Average Loss Per Incident ____________

Frequency of Occurrence x ____________

ALE Equals ____________

Security Measure(s) Being Considered ____________________

____

Initial Cost ____________

Annual Maintenance ____________

First Year Cost ____________

Life Expectancy *(Years)* ____________

Total Cost Over Life Expectancy ____________
(Initial Cost + Annual Maintenance x Life Expectancy)

Expected Reduction in ALE ____________
(Reduced Losses)

Return on Investment (ROI)

First Year ROI ____________
(Expected ALE Reduction – First Year Costs)

A. Total Reduction in ALE ____________
(ALE Reduction x Life Expectancy)

B. Total Cost of Security Measure ____________
(From Above)

Total Estimated ROI Over the Life of the Measure

(A – B) ____________

Fig. 4–2 ALE Worksheet

The intolerable SLO calculation is used when considering a potential loss that has a low probability of occurrence but a major consequence if it does occur. The potential consequence of an SLO can be computed or estimated.

What is the consequence of your main office building burning to the ground? Losses would include rental of other office space during the rebuilding, the cost of rebuilding, the loss of records and equipment, down time for all of the functions performed in the building, decrease in employee productivity, *etc*. Some of these consequences can be computed (*e.g.*, rebuilding); others would be estimated (*e.g.*, decrease in employee productivity). Once these costs are captured, subtract the total from the insurance that would cover the loss, and this is the SLO for this event. If the result is negative, you have insufficient insurance to cover this loss. How much would additional insurance cost? What other approaches could be used to mitigate the SLO? There are several alternatives to purchasing additional insurance:

- Distribute the risk by not putting all of your assets together. Is it possible to move some of these functions to other locations?
- Reduce the vulnerability to the risk by improving the fire protection system. Fire may not be the only risk if you are in an area prone to hurricanes, tornadoes, or earthquakes.
- Reduce the vulnerability of the loss by improving contingency planning and recovery capabilities. Back-up media could be stored off-site, and the utility could contract for a *cold site* as a backup to its network and computer systems.

The risk analysis committee or security managers have the responsibility to identify incidents with a low probability but high consequence and to compute the SLO for each. These recommendations are then submitted to management, which decides how much risk is acceptable and how unacceptable risk should be managed. The security director or committee then implements the risk-reduction approaches based upon the management decisions and directives.

Summary

A good deal of security money is spent on terrific systems and procedures that upgrade security without anyone first identifying the threat to the assets being protected. This is the *guards and gates* approach: "Spend money, and the asset is secure." Not necessarily!

Before you begin to design your security program, identify specific threats to your assets to determine which assets are critical. You cannot protect everything on the system. Spend your time, effort, and money protecting those facilities that are critical. Once these facilities are identified, assess their vulnerabilities. *Now* we're ready to begin evaluating and recommending risk reduction measures and security procedures.

Action Checklist

Risk equation: *R = Threat + Criticality/Vulnerability*

1. Identify the team that will conduct the risk analysis for your utility. A larger company will have a number of teams. Each team should use the same risk equation determinations and checklists. They should also use the same report format. Teams should include security and operations personnel. If security personnel are not available, consider inviting crime prevention officers from a local police agency to participate.
2. Identify the critical facilities and components on your system.
3. Conduct the threat analysis.
   a. List previous incidents and threats at each location. The utility should be keeping a log of these events.
   b. Discuss the workplace violence prevention model with managers and employees at each critical facility. List the concerns next to the appropriate component of the model.
   c. Evaluate the census data for your areas using the *fedstats* web site.
   d. Assess the crime situation in the area using the data available at the Bureau of Justice Statistics web site. Also consult local law enforcement.
   e. Determine if there are known extremist groups in the area using (1) a major search engine, (2) the Department of Justice web site, and (3) the site for the SPLC.
   f. Determine the weapons and tactics an adversary may use against your critical sites.
   g. Use the terrorist target selection criteria to assess the possibility that critical sites could be targeted by extremists.

4. Determine criticality.
   a. Consider all laws and regulations that potentially affect each critical facility.
   b. Use the criticality determining criteria to evaluate each facility.
   c. Apply the criticality determining criteria to the assets and components at each critical facility.
5. Assess vulnerability.
   a. Complete the FVDS, Appendix B.
   b. To which weapons and tactics is each critical point vulnerable?
   c. Diagram an ASD for each critical/vulnerable asset or component.
   d. Develop an EASI for each critical location.
6. Reduce risk.
   a. Consider the criteria that determine the need and utility of a security reduction measure. Apply these to the measures you recommend.
   b. Prepare a physical protection and vulnerability matrix for each critical site.
   c. Compute ALE worksheets as appropriate.
   d. Prepare SLO recommendations as appropriate.

Notes

1. Garcia, Mary Lynn, *The Design and Evaluation of Physical Protection Systems* (Woburn, MA: Butterworth-Heinmann, 2001), p. 272

2. *Ibid.*, p. 35

3. Scalingi, Paula, "Assuring Energy in an Interconnected World," presented at Safety & Security in the Electric Power Industry, Houston, March 25, 2002

4. Seger, Karl A., *The Anti-Terrorism Handbook* (Novato, CA: Presidio Press), 1990

5. Garcia, 2001

6. Caroll, Jami M., "A Metrics-based Approach to Certification and Accreditation," BTG, Inc., available from jcarroll@btg.com

5

Protecting Information

What Needs to be Protected?

We live in an information age. Never before has so much information been available to so many. Unfortunately, some of this available information increases risks to the utility's security and the security of its employees and customers.

The information that utilities possess needs to be protected to the greatest extent possible. This presents special challenges to government-owned utilities, where information is often available to the public under the state's "sunshine" laws.

An inquisitive patron entered the lobby of a utility and asked for the address of an old friend he was looking for. Because this is a municipal utility, the information was not protected and was provided. The person who was asking for the information was a "hit man" working for organized crime. The person he asked about was scheduled to be a witness against members of the crime group. The murderer used the information obtained at the utility to locate the witness and kill him.

Not all utilities have adjusted to the new competitive nature of the industry and fail to understand the value of information pertaining to their operations. Some requests for information are overt. In one case, a major investor-owned utility asked for all of the financial data from a large municipal system it was considering for an attempted acquisition. The employee who received the information request compiled the data, mailed it to the investor-owned utility and then, in passing, mentioned it to his supervisor.

Other attempts at compromising information are more covert. A substation engineer, who had developed a unique approach to designing substations for his utility, received a telephone call from another company asking if he would be available to assist them in designing several new substations. He would, of course, be paid a substantial consulting fee. As requested, he mailed his unique design plans to a post office box in another state. Several months later, he met the chief substation engineer from that utility at an industry meeting and asked about the person who requested the information and the status of the project. He learned that no one by that name worked for the company, nor did they have a post office box in the city where he sent the information. Some less-than-ethical-consultant is probably selling the engineer's concepts to utilities in other regions of the country.

There are at least five categories of information the utility should protect:

- proprietary information
- employee information
- customer information
- security-related information
- competitive information

Proprietary information protected by the utility will depend upon its ownership. In addition to Security and Exchange Commission (SEC) filings, investor-owned utilities must file regulatory reports and, in some cases, may have to file VAs of their critical facilities with government regulatory agencies. (Since the information on file at these agencies is often public record, are these VAs subject to disclosure to the public?) Confidential information related to finances, operations, and other proprietary information must also be protected.

As stated, government-owned utilities may have greater challenges in protecting information, even in a competitive environment. In some cases, information collectors claim that all information at these utilities (including employee and customer information) should be available. How much of this information is available through "open source requests" depends upon the law and court decisions in the jurisdictions in which you operate.

Since cooperative utilities are owned by their members, information regarding operations and finances is usually made available to the membership. Employee and membership information may not be available.

Regardless of the ownership structure, all personal employee information should be protected. A general employee list could become a target list for a disgruntled customer. If the list includes employee home addresses, the information becomes even more critical. Obviously, personnel and medical information is highly sensitive and should be protected.

Customer information is also sensitive. In some cases, courts and regulatory agencies have ruled that even customer usage records are privileged and should not be released without appropriate legal authority.

When an electric meter reader noticed an exceptionally high consumption at a house he suspected of being a marijuana growing operation, it was reported to the local police. The owner of the property was arrested and convicted on multiple charges. However, his conviction was dismissed on appeal when the appellate court ruled that the utility should not have released the customer's consumption records without a subpoena. The state regulatory agency also chastised the utility for releasing the information.

Once you complete a risk analysis at each of your critical facilities and have identified the threat, criticality, and vulnerabilities, because these documents include the countermeasures recommended to reduce the vulnerabilities identified, they become a blueprint for an adversary. This information must be protected. There are other security-related information targets an adversary will attempt to acquire; several examples are listed below.

- How many people work at a critical site, and what are their schedules?
- Do you have security guards at any sites, and are they armed?
- Where are your intrusion detection systems and security cameras located?
- What are the money-handling procedures at the utility?
- When are your cashiers most vulnerable?
- Are employees collecting cash payments in the field?
- What materials are stored in your warehouses?
- Any other information that would help plan a successful robbery, vandalism, or other crime against the utility.

During the past decade, the utility industry worldwide has become increasingly competitive. Different services (*e.g.*, electric and gas) compete for major energy users. Companies compete for national, regional, and local contracts. There have been a high number of acquisitions and mergers, some friendly and others hostile. This means that competitive intelligence has become a concern for many utilities.

Almost all competitive intelligence information is collected through open sources. These include newspapers, magazines, and journals, interviews with former and current employees, and information collected at industry meetings and trade shows. Annual reports and regulatory filings are important sources of information. The utility's web site often provides valuable competitive information. If competitive intelligence is a concern to your utility, keep a file on your utility so that you know what is available to others. The file should include information from a number of sources including the following:[1]

- Newspaper, magazine, and journal articles
- Brochures your utility publishes and distributes
- All of the information on your web site
- Regulator filings and related documents
- Trade shows and industry meetings
- Research and development sources

- Financial periodicals, including investment newsletters
- Information found using Internet search engines
- SEC filings including annual reports
- Industry directories
- Credit reporting services
- State corporate filings

In today's world, information is power. If you don't protect the critical information about and at your utility, you not only lose this power, but you may find yourself changing the logo on the door.

For more information on the threat of economic espionage, contact the National Counterintelligence Executive at (703) 874-8364. You can download information on the threat from their web site, *www.ncix.gov*.

Operations Security

Almost every security countermeasure you implement is predictable. The utility will use firewalls, password controls, and other hardware and software to protect its network. But these "systems approaches" are a standard in the information security industry, and the equipment is off-the-shelf. A determined adversary will learn more about your system protection than you may know and take advantage of this knowledge.

Systems used for access control, physical security, intrusion detection, and surveillance are all standard-issue. An adversary can learn the specifications for each system by reading the vendor's catalog or by going to their web site, and then look for weaknesses to exploit.

Most of these security systems will help you to achieve your objectives, but all of them are predictable. To make them *less* predictable (and more effective), you should introduce *operations security* into your program.

There are three primary objectives to an operations security (OPSEC) program

1. Identify information that needs to be protected, and install systems to protect it.
2. Introduce an element of unpredictability to all aspects of your security program.
3. Increase the level of awareness among employees and others who work at your facilities.

According to the Interagency OPSEC support staff (U.S. government), OPSEC is defined as follows:[2]

1. *A systematic, proven process by which a government, organization, or individual can identify, control, and protect generally unclassified information about an operation/activity and, thus, deny or mitigate an adversary's/competitor's ability to compromise or interrupt said operation/activity (NSC 1988).*
2. *OPSEC is a process of identifying critical information and subsequently analyzing friendly actions attendant to military operations and other activities to (a) identify those actions that can be observed by adversary intelligence systems, (b) determine indicators adversary intelligence systems might obtain that could be interpreted or pieced together to derive critical information in time to be useful to adversaries, and (c) select and execute measures that eliminate or reduce to an acceptable level the vulnerabilities of friendly actions to adversary exploitation (DOD JP 1994; JCS 1997).*

Your utility is not preparing for war or protecting national security secrets, but you should still have an OPSEC program to protect OPSEC indicators that telegraph important information to potential adversaries. For example, an armed courier service picks up your bank deposit every day at 3:30 P.M. This is an action that takes place in plain view of anyone watching. What does this indicator mean to a potential robber?

- You handle a significant amount of cash.
- You require a courier service, and that the best time to rob you is about 3:20 P.M., when the deposit bags are ready for pickup.

Other examples of OPSEC indicators may be found on your web site or on other web sites. Some utilities provide a system map on their web site identifying all of the critical facilities. Major transmission grids and interstate pipeline systems are found on web sites these utilities do not control.

There are five steps to developing an OPSEC program:[3]

1. Identify the critical information at your utility.
2. Conduct an analysis of the threat to this information.
3. Determine the vulnerability of this information to compromise.
4. Assess the risks associated with information compromise.
5. Apply appropriate countermeasures to protect this information.

There are a number of multiple-step approaches to developing OPSEC programs designed to protect national security-related information. We will use a simplified three-step process.[4]

The first step is to identify the OPSEC indicators that an adversary would seek to compromise. These can be identified during a brainstorming session in which operations and security personnel participate. Brainstorming rules apply:

- Everything that is suggested is written down on chart paper by the recorder.
- There is no ownership of the ideas suggested.
- There is no criticism of the ideas suggested.
- Participants are encouraged to present ideas as the facilitator goes around the room.
- If someone doesn't have an idea during their turn, they simply say, "Pass."

Once the ideas are listed, take a break. After the break, go back through the list. There will be repetition, additional ideas suggested, and some ideas will be discarded. Each of the ideas on your list is an OPSEC indicator and part of critical information you should protect (OPSEC Development Step 1).

Schedule the next brainstorming session for about a week later. This is so participants have the opportunity to think of OPSEC indicators that may not have been discussed during the first session. Add these to the list.

The objectives of the second brainstorming session are to:

- conduct an analysis of the threat to this information.
- determine the vulnerability of this information to compromise.
- identify the risk associated with information compromise.

Your chart paper has four columns. The first column lists the OPSEC indicators identified during the first session. The other columns are for the threat, vulnerability, and risk associated with those indicators (OPSEC Development Steps 2 through 4).

Be prepared for a longer brainstorming session. This should take at least a day. Again, follow the brainstorming session with a meeting to clean up the ideas presented, eliminate duplication, and present additional ideas. Take another week off.

You have now identified the critical information at your utility (OPSEC indicators), the threat and vulnerability of these indicators, and the risk associated with information compromise. During your third and final brainstorming session, identify countermeasures that could protect this information. During the brainstorming session, all of the ideas presented are recorded. Your OPSEC plan is developed during a meeting that takes place after this session. The plan is now ready for presentation to management and, once approved, ready for implementation and presentation to your employees (OPSEC Development Step 5).

It is important for all employees to understand they have a role in the OPSEC program. For instance, if your utility has an aggressive revenue-protection program (where utility thieves are caught and prosecuted), some of these prosecutions will be publicized in the local media. As a result, personnel working in the field may be asked about the program by customers. If asked how utility thieves steal, should your employees respond by saying that some electric meters run backward if placed in an inverted position, or that gas thieves simply bypass the meter? Hopefully not! These are OPSEC indicators that tell people how to steal your services. Instead, employees should inform these customers that stealing is against the law and that it can result in injury and death, but don't expect employees to properly provide this information unless you have trained them as part of your OPSEC program.

Here's an example of "bad OPSEC." In December 2001, a major cable television network interviewed managers from several major water utilities in the United States and asked them if terrorist attacks against their source supplies were a concern. The water utility managers said they were not concerned with attacks on their supplies but rather that attacks could involve the release of hazardous material using *backflow techniques*. They elaborated on their concerns and provided a basic idea of backflow. The cable network made this information available on its television network and on its web site. Great training information for extremist groups!

Another example of bad OPSEC. Almost a year after September 11, 2001, a major federally owned electric generation and transmission system provided a complete map of the system to its web site. The map details the transmission system and the location of all of its generation facilities, including three nuclear reactors. It provides information on each of the generation facilities and shows how electricity is generated at each of the nuclear sites. Many of these sites are hydro-generation dams. Weren't dams among the high-threat categories identified following the attacks on America?

Not all federally owned utilities display bad OPSEC. Another agency has removed the generation and transmission maps from its web site and now provides little information that could be used for targeting. This agency provides information on recent security breaches at its facilities but then details either how the responsible parties were identified, or it discusses the reward program that provides incentives for reporting those responsible.

Take a careful look at your utility's web site. Does it provide information that would be of use to an adversary targeting your system? Conduct a search at a major search engine. What information is available about your utility from other Internet sources? Try *www.yellowpages.com*, as an example. You may find the location of each of your facilities and maps to their locations. There are other sites on the Internet that may provide satellite photographs of your facilities. What more could a determined adversary ask for?

When developing your OPSEC countermeasures, consider the Three Ds of OPSEC:[5]

- Denial
- Disguise
- Deception

Denial includes those countermeasures that prevent an adversary from obtaining OPSEC indicator information. *Disguise* countermeasures are a process used to make the information harder to compromise. (Important computer files can be renamed to make them appear to be innocuous. Contract guards can be asked to arrive at the site in street clothes so that an observer cannot determine how many, if any, guards are on the site at a given time.) *Deception* is harder to employ in situations that do not involve national security or other critical threats. Here's an example: When nuclear materials or other sensitive materials are transported, neither the driver nor the security personnel know for certain that the movement actually contains the sensitive material. Sometimes it does and sometimes it doesn't. This makes it increasingly difficult for an adversary to plan a hijacking of a shipment even if they are working in collusion with an insider.

For more information on operations security, including instructional materials and videotapes, contact the Interagency OPSEC support staff at (443) 479-4677 or go to their web site at *www.ioss.gov*.

Computer and Network Security

This topic is more fully covered in chapter 6, but some comments are appropriate here.

Introduce the topic of computer or network security to some utility managers, and you will see their eyes glaze over as they begin to yawn. Computer and network security is the responsibility of the IT experts, right? Not really. IT professionals need the support of management and the cooperation of all other employees in the utility if they are going to be able to initiate and maintain effective computer and network security programs.

Computer and network security requires teamwork. Management, security, and IT professionals need each other's help to identify the threats, critical points in the system, and system vulnerabilities. They need to then work together to address the threats and vulnerabilities at these critical points.

Computer and network security crime is a serious problem. An investigative organization within the DOD targeted 8932 systems for an attempted intrusion. Of these, 7860 systems were penetrated. Only 390 of these detected the penetration, and, of these, only 19 reported them, which is required by federal regulation.

During a critical power crisis in a western state, hackers were able to penetrate the system controlling the transmission grid and stay *in* for several days. Most embarrassing was the fact that the intrusions were not detected until several days after the hackers had moved on. These hackers were obviously *script bunnies:* inept hackers who download software from the Internet and use it to intrude into systems but don't know what to do once they succeed.

According to a report released in July 2002, cyber attacks targeting utility systems are increasing. Riptech, Inc., a Virginia-based security firm, had experienced a 77% rise in cyber attacks on their power and energy clients in the previous six months. Some of these attacks were believed to be malicious hackers or industrial espionage, but a small number (1260 out of 180,000) originated in countries where terror groups were known to be concentrated, including Kuwait, Egypt, and Pakistan. Ronald Kick, director of the FBI's cyber crime division, said he is concerned that the nation's power grid may be "moving into the cross-hairs of cyber-terrorists." Potential targets include computer systems used to control the flow of gas, oil, and water through pipelines as well as the power grids.

Utilities have been the targets of other computer crimes, including employee fraud. A key person at a utility retirement organization was able to "retire" nonexistent employees and collect almost $250,000 before being caught. (The embezzler was nabbed when a retirement check for a nonexistent employee was returned to the office while the criminal was on vacation.) In a separate incident, a financial person at a utility embezzled approximately $1.25 million in slightly more than a year before being caught.

Less ambitious computer thieves learned they could access the customer billing files at several utilities between the time when meter readings were entered into the system and bills were computed. By lowering the meter readings, they lowered their bills and the bills of friends and neighbors.

Perhaps the biggest threat that computers and networks present in the workplace is misused time. If an employee spends four hours a day in the break room, disciplinary action should be taken, but not all utilities effectively monitor employee use of the network and Internet. (This is not a problem unique to the utility workplace. Consider that approximately 80% of the visits to Internet auction and pornographic sites occur during business hours.) In a small utility, one with about 100 employees, the manager became concerned that a key person in the utility was spending too much time on the Internet. An after-hours' check of the computer's browser files resulted in four single-spaced pages of sites recently visited. About 70% of these were pornographic. Several may have been child pornography.

Your utility should have a computer and Internet use policy describing acceptable and unacceptable uses of the system and possible disciplinary actions for misuse. It should include information on your utility's policy regarding e-mail. Examples of these policies are found in Appendix C.

Your employees, vendors, and others with inside access to your system are a much greater threat than hackers in some cases. In a recent survey of more than 1000 computer/network security professionals, 53% reported that current employees pose a greater threat to their company's technology than former employees or other outsiders. According to a federal government report, the average cost to an organization after a hacker attacks is $56,000, while the average cost of a malicious act by an insider is $2.7 million.

Computer and network security requires management, system, and technical solutions as well as the cooperation of everyone who has access to your systems.

Management solutions include:

- Establishing and enforcing computer and network use policies
- An emergency response capability to respond to viruses and other emergencies
- Risk management controls
- Audit and evaluation systems
- Contingency planning
- Accountability

System solutions include:

- Initiating and maintaining administrative controls
- Physical and environmental controls
- Information and data controls
- Software acquisition and development controls
- Backup and contingency planning

Technical solutions include:

- Network security to include protecting information, controlling and monitoring e-mail, and network access control
- Data encryptions including key management, message authentication, electronic certification, and public key cryptography
- Password controls (ensure they are changed frequently and not shared)

- Control access to important data and other information
- Establish systems of accountability for use

Smaller utilities rely on local or state law enforcement for assistance when an incident involving computers or network occurs. You should have a liaison established and a plan in place before an incident occurs. If you are the target of a computer/Internet crime, you must be prepared to work with investigators to document all aspects of the incident, collect evidence as appropriate, and provide technical advice. Larger organizations should have a Computer Security Incident Response Team (CSIRT). Advice in establishing a CSIRT is provided by the Computer Emergency Response Team (CERT) at the Carnegie Mellon Software Engineering Institute. According to CERT, the steps required to establish a CSIRT include the following:

- Obtain management support and buy-in
- Determine the CSIRT strategic plan
- Gather relevant information
- Design the CSIRT vision
- Communicate the CSIRT vision and operational plan
- Begin CSIRT implementation
- Announce the operational CSIRT
- Evaluate CSIRT effectiveness

For information on establishing a CSIRT, download *Creating a Computer Security Incident Response Team: A Process for Getting Started* from the CERT web site at *www.cert.org*. There are a number of other useful publications at this site, and the CERT publishes a periodic summary for anyone involved with computer and network security. Summaries are located at *www.cert.org/summaries/*.

Another computer security report that your utility should be receiving is *CyberNotes*, an on-line publication from the NIPC. This comprehensive report, published every two weeks, provides timely information on cyber vulnerabilities, hacker exploit scripts, hacker trends, virus information, and other critical infrastructure-related nest practices. CyberNotes is available on the web at *www.nipc.gov/cybernotes/cybernotes.htm.*

There are a number of excellent reports and newsletters on the web published by private sector sources. Two of these sources are:

- Computerworld, *http://computerworld.com*
- Trend Micro Weekly Virus Report, contact *VirusInfo@ trendmicro-newsletters.com*

Summary

Information is a valuable commodity. It must be protected. Whether it is a document or other hard copy or information in an information technology system, procedures must be in place to secure sensitive and potentially sensitive information. As the utility industry becomes increasingly competitive, protecting information becomes a vital necessity.

OPSEC is one of the approaches used to identify critical information and to establish programs and procedures to make it more difficult for an adversary to gain access. OPSEC also provides the basis for increasing employee awareness of these threats, and it enhances the effectiveness of your other security programs by making them less predictable.

Protecting information is not an option. It is a necessity.

Action Checklist

1. Identify information that should be protected. These include:
   a. Proprietary information
   b. Employee information
   c. Customer information
   d. Security-related information
   e. Competitive information

2. Develop a file that includes information on your utility. Make sure you include the 12 potential sources of information discussed in this chapter.

3. Consider developing an OPSEC program to help protect sensitive information. Use the three brainstorming sessions to complete the five steps to developing an OPSEC program:
   a. Session 1—identify the critical information at your utility
   b. Session 2—conduct an analysis of the threat to this information, its vulnerability, and the risk associated with compromise
   c. Session 3—identify appropriate countermeasures to help protect this information

4. Conduct a critical assessment of your utility's web site. Does it provide critical information that could be of use to an adversary?

5. Develop a team approach to utility security. Your approach should include:
   a. management solutions
   b. system solutions
   c. technical solutions

6. Prepare to respond to a computer security incident.
   a. Smaller utilities should liaison with law enforcement
   b. Larger companies should form and train a CSIRT

7. Subscribe to on-line computer security reports.
    a. *www.cert.org/summaries/*
    b. *www.nipc.gov/cybernotes/cybernotes.htm*
    c. *http://computerworld.com*
    d. *www.trendmicro.com/en/security/report/overview.htm*

Notes

1. Fuld, Leonard M., *Competitor Intelligence: How to Get It, How to Use It* (New York: John Wiley & Sons), 1985

2. *Glossary of OPSEC Terms*, Interagency OPSEC Support Staff, Greenbelt, MD, 1998, p. 27

3. *Operations Security: An Overview*, Interagency OPSEC Support Staff, Greenbelt, MD, 1996

4. Seger, Karl A., *The Antiterrorism Handbook* (Novato, CA: Presidio Press), 1990, pp. 108–112

5. *Ibid.*, p. 112

6

Protecting Networks and Computer Systems

Threats to Utility Networks and Computers

In an article published in July 2002, a Virginia-based security firm reported a 77% increase in the number of cyber attacks on its power and energy clients in the first six months of 2002 over the previous year. In the same article, the head of the FBI's cyber crime division expressed concern "that the nation's power grid may be moving into the cross-hairs of cyber-terrorists."[1]

Riptech, Inc. says that in the first six months of 2002, 14 of its 20 energy clients suffered cyber attacks that would have disrupted the company networks if they had not been immediately detected. Of the 180,000 attacks analyzed in the report, 70% originated in North American or Europe. However, a small number of attacks originated in countries where terrorists groups are known to be concentrated.[2]

As electric utilities become more reliant on standardized network protocols and web-based communications, the vulnerabilities of these systems increases substantially. A hacker, be it a vandal, cyber-terrorist, or thief, intruding on a utility's system via a network or dial-in connection, could gain access to the business systems, including customer credit card or other identity information. The intruder may be able to open a back door to the power system, exposing the system to degradation and a power outage. Increased public access to transmission system data mandated by FERC Orders 888/889, coupled with the widespread dissemination of hacker tools, has significantly increased this threat.[3]

The cyber threat to power systems is not new. In May 1995, following a briefing from the National Security Agency (NSA), the National Security Telecommunications Advisory Committee (NSTAC) formed the Information Assurance Task Force (IATF) to work closely with the U.S. government to identify critical national infrastructures and to identify threats to those infrastructures. The committee published an extensive report that detailed threats to generation, transmission, and distribution systems.[4]

As the world becomes connected via the Internet, a new type of vandalism is occurring—"hacktivism." *Hacktivists* claim to be hackers who target specific networks, companies, or government agencies to promote their cause. Hacktivists include militant environmentalists, anti-utility groups, and anti-globalization extremists. Like other hacking groups, they have their own on-line publications and communications networks (see *www.thehacktivist.com*).

The ultimate hacktivists align themselves with terrorists and other militant organizations. There are at least three major alliances of hacker groups targeting U.S., Israeli, and Indian interests:[5]

1. ***UNIX Security Guards*** Anti-Israeli alliance formed by four groups in May 2002: Egyptian Fighter, rD, Inkubus, and ShellCode. This alliance was responsible for 87 digital attacks in its first two months.
2. ***World's Fantabulous Defacers*** Pakistani alliance of 12 member groups that opposes India's presence in Kashmir. The alliance also has an anti-U.S. and anti-U.K. agenda. The alliance was responsible for attacking 445 sites between November 2000 and July 2002.

3. ***Anti-India Crew*** A Pakistani alliance formed in July 2001. The alliance is anti-India, anti-U.S., and anti-U.K. It carried out 422 attacks between July 2001 and July 2002.

Hacker intrusions into utility networks are not the only threat to computers and information. What about the possibility of losing a computer that stores important information? The manager of a Gulf gas station was surprised to find two computers in the trash bin behind his station in early November 2002. When he plugged them in, he found they both worked. What surprised him was that the computers contained photographs and information on nuclear generating plants. The files included photographs of a control room and home telephone numbers of plant staff.[6]

Hackers and hacktivists are not even the greatest threats to utilities computers and networks. You must also consider threats from employees and others with authorized access to your system. The three most common computer crimes are:

1. Unauthorized use of the system or network
2. Theft of software or using more copies of a program than the license permits
3. Use of computers to steal other assets such as money or physical assets

Unauthorized use of the system includes employees who spend much of their time accessing web auction sites or other "unauthorized" web sites and hackers who gain access to the system and use it to store data stolen from another web location or as a gateway to the Internet. One utility was experiencing a number of attacks from a local college. Some students found a method for gaining access to the utility's network and were then using external connections to hack into other computers and networks, accessing gambling and pornographic sites. At another utility, several employees were running home businesses on the company computer. Their justification was that one of the managers used the system for personal use (he completed his master's thesis on the company computer system). So, why shouldn't they? Shouldn't they have the same opportunity?

Employees who download proprietary software, or software that was purchased for company use, are stealing. This is no different than if they were stealing tools off a truck or cash out of a cash drawer, but they don't always see it that way. There should be a policy in place that prohibits downloading of software without authorization, and it should be enforced. Also, ensure that when you purchase software with a license for X number of machines that you don't expand its use to X-plus without contacting the vendor for the additional licenses. This is another form of software theft.

In some cases, utility employees have used computer systems to steal small amounts of money or assets. Employees at two different utilities learned that they could access the main computer between the time that meter readings were entered into the system and when these readings were converted to dollars and cents for billing. They could not only *read* to the system, they could *write* to the system as well. They waited until their meter readings were entered and then changed the number of kilowatt-hours before the bills were printed. They are now ex-employees.

At another utility, a trusted employee was able to steal more than $1 million in slightly more than a year before he was detected.

The average fraud against a business where a computer is not involved results in a loss of $20,000. The average loss where a computer is used is $500,000.

Vulnerabilities

The NSTAC-IATF electric risk assessment identified three major vulnerable targets: control center vulnerabilities, substation vulnerabilities, and communication vulnerabilities.

The concern, according to the report, is that an electronic intruder could access the control center through several different interfaces including the following:

- Links to the corporate information system
- Links to other utilities or power pools

- Links to supporting vendors
- Remote maintenance and administration ports

The trend in the past decade has been to move from isolated, mainframe-based operating systems to off-the-shelf, commercially developed network systems. The move to standard vendor products, based on distributed client/server technology, makes the system more predictable and more easily compromised. There is also an increasing trend to use public networks to interconnect corporate networks with other utilities, vendors, and the Internet. This is a cost savings, but it increases the potential vulnerability of unauthorized access.

In an effort to provider better service to customers and to reduce staffing requirements, substations have moved from manual systems to a variety of intelligent electronic devices, including digital programmable breakers, switches, and relays. RTUs collect data for the control center and operate as a clearinghouse for control signals to transmission and distribution equipment. Some RTUs can be accessed through a dialup modem without proper protection, which makes them vulnerable to hackers.

As stated in the NSTAC-IATF report, utilities rely on a mix of private microwave radio, private fiber, and the public networks for communications among control system elements. Any of these components could be exploited by an electronic attack. A coordinated cyber attack on the communications infrastructure, in conjunction with an attack on the electric power control system, could result in what one utility official characterized as a "nightmare scenario."[7]

An attack on a utility's SCADA system could be disastrous. SCADA systems control the grid though which electric power is distributed from power production stations across the network of high-voltage transmission lines and ultimately to the customer through the distribution system. They are essential to effective energy and system management. As a result of the changes in systems that have already been discussed (*e.g.*, move to client-based systems, and use of public network communications), many SCADA systems may be at risk. If the utility is using a web-based application, the protections to the system should include the key features outlined below.[8]

- Installs and deploys easily to accommodate a wide range of technical expertise across organizations
- Self-configures to applications to account for undocumented legacy code and outdated deployments
- Accounts for cross-site/application vulnerabilities that could allow hackers to use enterprise web applications at an entry point
- Employs a positive security model to protect against unknown attack structures and changing application environments
- Is compliant with various government regulations and requirements

An assessment of the electric transmission system in the United States following the events of September 11, 2001, determined that the grid could be extremely vulnerable. Some load-based power plants are in remote areas close to fuel sources and away from environmental concerns; as a result, there may be limited security and increased vulnerability. Transmission lines run across thousands of miles of remote, unprotected areas. To address threats to transmission systems, it is recommended that systems be planned with double contingency (N-2) criteria. The system should remain stable during the outage of any two lines or substations anywhere on the system.[9]

NERC and Cyber Security

The FBI and the NIPC have been addressing cyber security threats for several years. In an effort to work closely with the private sector, the FBI developed the InfraGuard program and has InfraGuard coordinators in each of its 56 field offices. The program offers four basic services to its members: secure and public web sites, an alert and incident reporting network, local chapter activities, and a help desk. For more information on InfraGuard, contact the FBI office nearest you. The program expected 4000 new members in 2002.[10]

The NIPC established the ISAC to provide a two-way information sharing process and increased security for the nation's infrastructure. The energy unit is one of the sectors represented within the NIPC; the NERC serves as the electric power ISAC. (NERC physical security guidelines were discussed in a previous chapter.) NERC has also published cyber response guidelines and a cyber alert threat system. The NERC ISAC analyzes threat information from law enforcement sources and from within the electric utility industry, issuing alerts as appropriate. Similar to the physical threat alert guidelines, the cyber threat guidelines are established at five levels, using the Homeland Security threat alert level categories.

Threatcon ES-Cyber-Green (Low) applies when there is no known threat of cyber attack or only a general concern about hacker activity that warrants routine security procedures. Any cyber security measures applied should be maintainable indefinitely and without adverse impact to business or expenses. This may be equivalent to normal daily conditions.

Threatcon ES-Cyber-Blue (Guarded) applies when there is a general threat of increased cyber (hacker intrusions, viruses, etc.) activity with no specific threat directed toward the electric industry. Additional cyber security measures may be necessary, and, if initiated, they should be maintainable for an indefinite period of time with minimum impact on normal business or expenses.

Threatcon ES-Cyber-Yellow (Elevated) applies when a general threat exists of disruptive cyber activity directed against the electric industry. Implementation of additional cyber security measures is expected. Such measures are anticipated to last for an indefinite period of time.

Threatcon ES-Cyber-Orange (High) applies when a credible threat exists of disruptive cyber activity directed against the electric industry. Additional cyber security measures have been implemented. Business entities need to be aware that corporate resources will be required above and beyond those required for normal business or expenses.

Threatcon ES-Cyber-Red (Severe) applies when an incident occurs or credible intelligence information is received by the electric industry indicating a disruptive cyber attack against the electric industry is imminent or has occurred. This condition may apply as a result of an incident in North America, outside of the electricity sector. Maximum cyber security measures are necessary. Implementation of such measures could cause hardship on personnel and seriously impact facility business and security activities.

The NERC *Threat Alert System and Cyber Response Guidelines for the Electricity Sector, Version 2.0* (Oct. 8, 2002) are included as Appendix B. Your utility should develop a cyber threat alert response plan just as you developed a physical alert response plan. You will find additional security guidelines and other information regarding threats to the electric industry at the NERC web site, *http://www.nerc.com.*

CERT and NSA

While most of the NSA operates in a highly classified environment, the Information Assurance Directorate's mission of the NSA is to help provide security information and standards for the private sector. These include security recommendation guides for:

- Windows 2000
- Windows NT
- Cisco Router
- E-mail and executable content guides

The guides are posted on the NSA web site and include ".INF" files, configuration guides, and supporting documents. The web site is located at *http://www.nsa.gov/isso/index.html.*

The CERT issues advisories and alerts on cyber threats. The products offered by the CERT are found at *http://www.cert.org*. One of the most important products is not for IT professionals but for managers, security directors, and others who are concerned about network security. *The Common Sense Guide for Senior Managers* was developed by the Internet Security Alliance (ISAlliance) and is available at the CERT web site. This guide was developed by the Best Practices Working Group of the ISAlliance and addresses 10 of the recommended highest priority security practices for today's operational systems. These include:

1. ***General management*** Managers throughout the organization consider information security a normal part of their responsibility and the responsibility of every employee.

2. ***Policy*** Develop, deploy, review, and enforce security policies that satisfy business objectives.

3. ***Risk management*** Periodically conduct an information security risk evaluation that identifies critical information assets (*e.g.*, systems, networks, and data), threats to critical assets, asset vulnerabilities, and risks.

4. ***Security architecture and design*** Generate, implement, and maintain enterprise- (or site-) wide security architecture, based on satisfying business objectives and protecting the most critical information assets.

5.1. ***Users issues: accountability and training*** Establish accountability for user actions, train for accountability, and enforce it in organizational policies and procedures. Users include all those who have active accounts (employees, partners, suppliers, and vendors).

5.2. ***User issues: adequate expertise*** Ensure there is adequate in-house expertise or explicitly outsourced expertise for all supported technologies (*e.g.*, host and network operating systems, routers, firewalls, monitoring tools, and applications software), including the secure operation of those technologies.

6.1. ***System and network management: access control*** Establish a range of security controls to protect assets residing on systems and networks.

6.2. ***System and network management: software integrity*** Regularly verify the integrity of the software.

6.3. ***System and network management: secure asset configuration*** Provide procedures and mechanisms to ensure the secure configuration of all deployed assets throughout their life cycle of installation, operation, maintenance, and retirement.

6.4. ***System and network management: backups*** Mandate a regular schedule of backups for both software and data.

7.1. ***Authentication and authorization: users*** Implement and maintain appropriate mechanisms for user authentication and authorization when using network access from inside and outside the organization. Ensure that these are consistent with policies, procedures, roles, and levels of restricted access required for specific assets.

7.2. ***Authentication and authorization: remote and third parties*** Protect critical assets while providing network access to users working remotely and to third parties such as contractors and service providers. Use network-, system-, file-, and application-level access controls, and restrict access to authorized times and tasks, as required.

8. ***Monitor and audit*** Use appropriate monitoring, auditing, and inspection facilities, and assign responsibility for reporting, evaluating, and responding to system and network events and conditions.

9. ***Physical security*** Control physical access to information and IT services and resources.

10. ***Continuity planning and disaster recovery*** Design business continuity and disaster recovery plans for critical assets, and ensure that they are periodically tested and found effective.

Each of the best practices area discussions includes specific questions for enterprise leaders, senior managers, and oversight boards, as well as a list of references.

Summary

Since one of the major threats to networks and computers is from insiders, your utility should monitor employee web and system use. There are a number of security systems that accomplish these objectives. Monitoring employee use will reduce visits to unauthorized sites and the personal use of the system. It may also reduce the utility's liability in the event an employee or other insider uses the system for illegal purposes including sexual harassment or accessing child pornography web sites. Some employees consider monitoring an invasion of privacy, but the results of a 2001 study found that more than one-third of the 40 million employees on-line at work in the United States have their e-mail and Internet use at work under continuous surveillance. There are five steps that will help to minimize misunderstandings regarding monitoring of "at work" use of the system and network access.[11]

- Disclose your plans in advance and encourage employee feedback.
- Have clear guidelines on what behavior is not acceptable.
- Be respectful of employee needs and time.
- Strike a reasonable balance between security and privacy.
- Hire people you can trust.

Utility computer and network security must go beyond the security considerations of the company or the enterprise; it must consider threats to the infrastructure of the country. In most cases, when a business network fails, it only impacts that business. When a utility's system is compromised and fails, the results can be a significant impact on the community or the region.

Computer and network security is not the sole responsibility of your IT staff. Managers and security professionals should obtain a copy of the ISAlliance *Common Sense Guide for Senior Managers* from the CERT web site. As stated in the guide, information security must be the responsibility of every employee if it is going to exist.

Action Checklist

1. Management and IT specialists should discuss the three most common computer crimes and the technology and polices in place at your utility to address these concerns.
   a. Unauthorized use of the system or network.
   b. Theft of software or using more copies of a program than the license permits.
   c. Use of computers to steal other assets such as money or physical assets.
2. Determine potential vulnerabilities at your utility that are discussed in the NSTAC-IATF power industry risk assessment report.
   a. Control center vulnerabilities.
      1) Links to corporate information system.
      2) Links to other utilities or power pools.
      3) Links to supporting vendors.
      4) Remote maintenance and administration ports.
   b. Substation vulnerabilities including RTUs.
   c. Communication system vulnerabilities.
   d. SCADA system security.
3. Develop a security alert system in accordance with the NERC cyber response guidelines and threat alert system.
   a. Access the NERC guidelines at *http://www.nerc.com.*
   b. Telephone your local FBI office and ask about the InfraGuard program.
   c. Review the appropriate security recommendation guides from the NSA web site, *http://www.nsa.gov/isso/index.html.*
4. Implement a utility-wide cyber security program.

a. Managers and IT professionals should review and discuss the *Common Sense Guide for Senior Managers* found at the CERT web site, *http://www.cert.org*.

b. Implement a system for monitoring employee use of the network and Internet. Involve the workforce in the development of policies and procedures regarding network and Internet use.

Notes

1. "Cyber Attacks on U.S. Power Companies on Rise: Report," *Xinhuua News Agency*, July 8, 2002
2. *Ibid*
3. "United States: SEL and Utilities Work to Safeguard Electric-Power Grid," *Transmission and Distribution World*, January 1, 2002
4. NSTAC, *National Security Telecommunications Advisory Committee Information Assurance Task Force Electric Power Risk Assessment*, *http://www.aci.net/kalliste/electric.htm*
5. "MI2G: Pro-Islamic Hacker Groups Join Forces Globally," M2 *PressWIRE*, June 18, 2002
6. "Vermont Yankee Computers Found in Trash Bin," *Associated Press*, November 11, 2002
7. NSTAC
8. Chauhan, Abhishek, "Protecting Web-Exposed SCADA Systems Statum 8's APS Secures Power Industry, Applications in Real Time," *Statum8 Networks*, October 2002
9. Beehler, P.E., Michael E., "Transmission System Security After September 11th," Burns and McDonnell, *http://images.tdworld expo.com/files/1104/OUT4%20-%20Beehler.pdf*
10. Dick, Ronald L., *Cyber Terrorism and Infrastructure Protection, statement before the House Committee on Governmental Reform, Government Efficiency, Financial Management and Intergovernmental Relations Subcommittee*, July 24, 2002
11. Enbysk, Monte, *Should You Monitor Your Employee's Web Use?* Microsoft Corporation, 2002, *http://www.bcentral.com/articles/enbysk/156.asp*

7

Physical Security and Access Control

Imagine a secure state-of-the-art courthouse in a major city. Its design includes an underground passage from the jail to the courthouse. An internal hall system ensures prisoners being brought to the facility for trial cannot access general public areas of the four-story building. Surveillance cameras identify and record anyone approaching the internal garage where judges and other law enforcement officials park their vehicles. A panic alarm under each judge's bench means that if something happens in his or her courtroom, the deputies providing security in the building are notified. This is the way the building was planned and built. Good security? Not really.

As with most government buildings, there is no smoking in this facility. Employees who do smoke learned that the quickest way out of the building during their breaks is through the prisoner corridor system. These doors are propped open to facilitate the needs of the smokers and negate the secure movement of the prisoners.

Someone became concerned about the security of the judges' vehicles after they were parked in the garage, so they realigned the cameras to face inward. Now we have a good view of the judges' cars when they are parked but, unfortunately, no longer have a view of people approaching the garage. If a person chooses to roll a hand grenade under a car from the door, we get to see the grenade explode but not the person who threw it into the garage.

And the panic alarm system? We saved money here. Rather than employing a separate enunciator for each panic alarm (a separate alarm for each courtroom), the same bell goes off regardless of which judge activates the system. The deputies responding have to try to figure out which courtroom has a security alert.

The planned or intended use of a security measure is not always the same as its reality. Plans and intentions are a good beginning, but it is the reality that determines your vulnerabilities. In this chapter, we will discuss approaches to physical security and access control. In the later part of the chapter, we will discuss several approaches for maintaining these systems and ensuring that intention equals reality.

Physical security and access-control programs are essential to your utility's security. Effective programs are a psychological deterrent to criminals and other adversaries. They decrease the need for security personnel. They help to channel people through controlled areas and to confuse intruders. However, there is no absolute physical security. All physical security is measured in *penetration time*, the time it takes for a normal adversary, using readily available tools, to break through the barriers between the adversary and the target. Some of the key components of physical security are:

- protective barriers, including clear areas, fencing, perimeter entrances, and exits
- protective lighting to discourage unauthorized entry and to identify intruders
- structural security including building construction, doors, and windows
- access controls, including:
  - lock and key controls
  - pass-through systems, including employee badges, proximity cards, and biometric identification
  - visitor control systems

- employee control systems
- surveillance-camera monitoring of key areas
- use of security guards

Dealing with a potential threat from terrorists and other extremists means that a number of specific areas require special attention. Start with labels on buildings and parking spaces. Extremists go after specific targets, including specific areas within a facility and key individuals. Labeling targets doesn't make sense.

When possible, add barriers around high-threat areas. An additional chain link fence or other barrier adds to the penetration time and can help to achieve this goal. There is a balance that must be achieved. If the barrier limits line of sight to the critical area from the road or other vantage point, it may be an asset to the adversary. Once in, detection from the outside of the facility will be difficult.

Intrusion detection systems should be used at high-threat locations, but the right system at the right location must be used, otherwise there will be an unacceptable number of nuisance alarms. Considerations for alarm and intrusion detection systems are discussed in the "bells and whistles" section of this chapter.

If extremists are a concern, you may have to maintain personnel at facilities that are not normally manned. Do you really want to put unarmed operational personnel at these sites when there is a high threat? The first question to ask is if this is really a critical site that needs to be manned? The second question is what are the responsibilities of the persons manning the sites? Do we want them to report intruders or stop them? Do we want to accept the liability of having an operations employee confronting a crusader, criminal, or crazy? Serious personnel security and liability issues arise when critical facilities are manned solely by operations personnel during high threat periods.

In one case, an electric utility put an intrusion detection system on a warehouse at a location different from its main facility. System alarms were received at the 24-hour dispatch office at the main facility. When an alarm was received, an on-call employee was sent to investigate. Does it make sense to send an employee to confront a possible intruder in the middle of the night? The procedure has been changed so that the employee does not approach the property until joined by a patrol officer. The patrol officer enters the property first.

What Are We Protecting?

As we begin to assess physical security and access-control needs, let's first consider what we are protecting. These considerations include the physical property and the facility operations.

Examine a plot or building plan (you might use a fire evacuation plan if the asset is a single building). What are the accesses and egresses to the facility? Are there other potential points of access? Where are the critical areas and how would these be accessed once inside the facility? Identify the physical security and access controls in place.

Now consider the operations within the facility. Any area where money is handled is a critical area, but there are other critical areas, as well. Areas where your computers, servers, mainframes, telecommunication, and other technical equipment are maintained are critical. Access should be secured. Remove signs that say "computer room" or other obvious indicators. Do the same with SCADA and crisis management centers. These should have controlled access, and no signs should designate the activities conducted there. In one utility, the SCADA/crisis management center is located off the break room where employees and vendors working for the utility congregate. A disgruntled employee of either the utility or a vendor could have easy access to the room and, if he knew what he was doing, create havoc for the system. The door used to be left open. It is now secured.

Depending upon the activities at any given location, you may have regulatory requirements, safety considerations, or legal issues to consider. Physical security and access control cannot negate these considerations. A balance must be achieved between issues such as restricting access into the building while providing immediate egress through the same doors in the event of a fire or other emergency. There are a number of access-control systems designed to meet this need. Other needs may be more challenging.

Physical security and access control are components of the risk management process. As discussed in chapter 4, identifying the risks and recommending procedures for controlling them is a security function, while determining the degree of risk that is acceptable and the procedures and countermeasures used to mitigate it is management's responsibility. Be prepared to be creative in developing physical security and access-control procedures in case you don't get all of the hardware you request.

Be prepared also to consider access-control measures at different threat levels. Access gates and doors that are open during normal operations may be secured at increased threat levels. For example, assume that there are two public access doors and three employee access doors to your facility during normal operations. The NIPC issues a high-level threat alert. You may now limit customer access to one door and employee access to a separate door. How will you secure the other doors? Who is responsible for ensuring that these doors are secured and that access is limited? (Does intention equal reality?)

Look at building or site plans and label all critical areas and the physical security and access controls that limit access to them. Now determine if additional barriers are needed and, if so, what you need and how you will sell the need to management and to the personnel who work at the facility. Remember the courthouse. The intentions were to secure the building. The physical security, access control, and other systems were in place, but the intentions did not equal reality because the need for the security systems was not communicated to the people who worked there.

The Basics of Physical Security: Some Specifics

There are recommended physical security standards that enhance your security program. These standards vary depending upon the source and may have to be modified for particular situations. For instance, there should be a clear area of 20 feet on either side of a barrier fence, but this may not be possible if the fence abuts a cliff. Or the facility may be on the property line of a farmer growing corn; are you going to pay him for his losses if he gives up 20 feet of growing space for your security? There should be continuous lighting on the outer perimeter of the fence, but if the facility is in a residential neighborhood, there will be complaints regarding the lighting.

Use the tables in this chapter as checklists for each of the areas discussed, but consider them starting points. Work with your security personnel to make the checklists site-specific. You may also be required to meet certain regulatory or legal standards; if so, incorporate them into your checklist. The objective is to develop a set of physical security and access-control checklists that are site-specific to your needs. These checklists will be used to conduct an annual security review at each facility.

Use an outside-in approach when conducting the survey, *i.e.*, begin with the area around the facility and work your way to the building or critical asset. Start with the protective barriers (Table 7–1).

One of the least expensive countermeasures is protective lighting. It discourages unauthorized entry and simplifies intruder identification. There are several different categories of protective lighting:

- stationary luminary/continuous lighting (overlapping cones of light often used in parking lots)
- controlled lighting (limits the illumination to a specific area)
- area lighting (targeting shadow areas)
- surface lighting (used to illuminate building surfaces)
- standby lighting (illuminated at increased threat levels or as a result of a suspected intrusion)
- moveable lighting (mounted on wheels or trailers)
- emergency lighting (emergency lights that activate in public buildings when power is lost)

Use Table 7–2 as the basis for your lighting inspection. Again, expand on this checklist to ensure it is site-specific. This should be obvious, but you need to inspect the security lighting at night.

Table 7–1 Protective Barrier Checklist

BARRIER	COMMENTS
PROTECTIVE FENCING (CHAIN LINK)	
Fence 6 to 7 feet high	
Wire is 9-gauge or heavier	
Opening not more than 2 inches	
Clear zone (20 feet on each side of fence)	
Check base for washout areas	
TOP GUARD FENCING	
Overhanging barbed wire	
Face outward at 45 degrees	
Minimum of three strands	
PERIMETER ENTRANCES AND EXITS	
Active entrances and exits are monitored	
Inactive entrances and exits secured	
Contingency plan in place to limit access during increased threat levels	

A utility building a new warehouse saved money by purchasing bay doors made of a plastic-like material. The doors looked nice until the weekend when someone cut through the back fence, kicked a hole in a bay door, and left with thousands of dollars of copper conductor. Structural security is not an area where the utility should buy the cheapest material.

Table 7–2 Security Lighting Checklist

PROTECTIVE LIGHTING	COMMENTS
ESSENTIAL PLANNING STEPS	
Maintenance requirements	
Impact of power interruption	
Impact of weather and climate	
Threat of fluctuating or erratic voltage	
Ledge to replace lighting at 80%of expected life	
PERIMETER LIGHTING	
Glare should not blind employees	
Lighting should not silhouette employees	
Employees control system	
BUILDING ENTRANCES AND EXITS	
Minimum 1.0 foot candle lighting	
Cover 8 feet in all directions	
Cover night deposit and other critical areas	
GENERAL CONSIDERATIONS	
All switches are inside building(s)	
Contingency plan in place to increase protective lighting during increased threat levels	

Table 7–3 begins to assess construction materials, doors, windows, and other structural considerations. Work with your security personnel or the crime prevention unit from the local police department to enlarge this list to make it site specific.

Table 7–3 Structural Security Checklist

STRUCTURAL CONSIDERATION	COMMENT
CONSTRUCTION CONSIDERATIONS	
Exterior and interior walls	
Windows and entrances	
DOORS	
Frame *(minimum 2"-thick wood or metal, rabbeted jamb to withstand spreading)*	
Door *(2"-thick solid wood or 16-gauge sheet steel)*	
Lighting *(Minimum 60-watt illumination)*	
Locks *(dead bolt or dead latch with 1" throw and anti-wrenching collar)*	
WINDOWS	
Burglary-resistant materials	
Secure areas *($^1/_2$" steel bars 5" apart)*	
Secure windows to prevent lifting	
OTHER CONSIDERATIONS	
Secure other openings larger than 96" with bars or screen	
Secure skylights with bars not more than 5" apart or with mesh screen	
Elevators: CCTV and continuous lighting	
Elevators: card-controlled access after hours	
Hatchways secured with slide bars	

Bells and Whistles

Your physical security system is not complete unless it has some bells and whistles. These include:

- Intrusion detection systems
- Alarm systems
- Surveillance cameras

These systems must operate in tandem. The intrusion detection system activates the alarm, and surveillance cameras investigate and record the suspected unauthorized entry. The entry control system will also activate the alarm if someone tries to subvert it. The individual bells and whistles must be designed to operate as an integrated system.

Intrusion detection

Intrusion detection is defined as:

> *...the detection of a person or vehicle attempting to gain unauthorized entry into an area that is being protected by someone who is able to authorize or initiate an appropriate response.*[1]

There are two basic types of intrusion alarm systems—exterior and interior. With both types of systems, the probability of detection, vulnerability of being defeated, and the potential nuisance alarm rate must be considered. Less expensive exterior intrusion detection systems may be especially susceptible to nuisance alarms since they may be triggered by vegetation, wildlife, or the weather.

Intrusion detection sensors are either active or passive. An *active sensor* emits some type of signal or energy and detects a change in the area in which it is active. Microwave sensors are active in that they establish an energy field using energy in the high electromagnetic spectrum.

The sensor is constantly transmitting and receiving signals. When there is a difference between the signal emitted and the signal received, the alarm is activated.

Passive sensors detect an action emitted by the target. A passive sensor on a door or window doesn't transmit energy, but when the intruder attempts to open the door or window, the sensor is activated.

Some sensors are intentionally visible while others are hidden. Visible sensors are used to deter intruders. However, they also let the potential intruder know that you have a system, and the determined professional or extremist may then try to figure a way to defeat it. The best system will include visible and covert sensors. The petty crook will be deterred by the visible sensors; the professional will be detected by the covert sensors.

The selection, installation, and maintenance of exterior intrusion detection systems should be undertaken by specialists. There are three different types of sensors to consider, and each requires expertise in the selection and installation if the end product is going to be reliable and not result in an unacceptable number of nuisance alarms:

- Buried-line sensors
- Fence-associated sensors
- Freestanding sensors

There are four types of buried-line sensors. Each depends upon different sensing phenomena:

- Pressure or seismic
- Magnetic field
- Ported coaxial sensor
- Fiber-optic sensors

Pressure or *seismic sensors*, buried in the ground, are passive. They respond to a disturbance in the soil caused by an intruder. Pressure sensors are sensitive to lower-frequency pressure waves in the soil, and seismic sensors are sensitive to higher frequency vibrations.

Buried *magnetic field sensors* are also passive and covert. They are underground and respond to changes in the local magnetic field caused by the nearby movement of metallic material. They are effective for detecting vehicles or intruders with weapons.

Ported coaxial cable sensors are active and covert. They respond to the motion of a material with a high dielectric constant or a high connectivity near the cables. These materials include the human body and metal vehicles.

Fiber-optic sensors allow light to travel through transparent fibers. If there is any change in the diffraction pattern or the light intensity at the end of the fiber, it will be detected. An intruder stepping into an area where fiber-optic cable is woven into a pattern beneath the ground will be detected.

There are three types of fence-associated sensors:

- Fence-disturbance sensors
- Sensor fences
- Electric-field or capacitance sensors

Fence-disturbance sensors are passive and visible. They can be installed on chain link fences and detect motion or shock. Because these sensors respond to all mechanical disturbances of the fence, wind, debris, rain, hail, and other natural occurrences can trigger alarms. Signs on the fence, when affected by wind, can also result in nuisance alarms.

Sensor fences are passive and visible. They make use of the transducer elements to form a fence. They are designed primarily to detect climbing or cutting on the fence. These sensors tend to be less susceptible to nuisance alarms provided they are correctly installed.

Electric field or *capacitance sensors* are active and visible. They detect changes in capacitance coupling among a set of wires attached to, but electrically isolated from, the fence. The sensitivity field can be adjusted to extend up to one meter beyond the wire or plane of wires. These sensors are susceptible to nuisance alarms resulting from lightning, rain, fence motion, and small animals. Ice storms can result in extensive damage to these systems.

Freestanding sensors in use today include:

- Active and passive infrared
- Biostatic and monostatic microwave
- Video motion detection

Active infrared sensors are active, visible, and line-of-sight. Many systems use a single beam, but multiple-beam systems are available for high-threat security systems. Vegetation and animals can result in nuisance alarms, and the sensors have a narrow plane of detection.

Passive infrared sensors detect thermal energy emitted by the warmth of the human body. They should be mounted so that the motion of the intruder will cross the most sensitive line of sight of the sensor. They are most sensitive when the background temperature is different from the intruder's.

Biostatic microwave sensors are active and visible. The typical installation is two identical microwave antennas at opposite ends of a detection zone. One of the antennas is connected to a transmitter and the other to a receiver. The sensor detects a change in this signal. These are often used to detect a person crawling through the protected zone. There are two important considerations to this application. The ground surface between the transmitter and receiver must be flat, or there will be areas where an intruder will not be detected. In addition, no detection exists in the first few yards in front of the antennas.

In *monostatic microwave sensors*, the transmitter and receiver are in the same unit. Radio-frequency energy is pulsed from the transmitter, and the receiver detects a change in the reflected signal caused by an intruder.

Video motion detectors are passive and covert. They sense a change in the video signal level in the protected zone. The video cameras can be used for detection and recording of an event when an alarm is triggered. Since these sensors are triggered by any movement in the zone, they are subject to nuisance alarms from small animals, cloud shadows, precipitation, and blowing debris.

As with external intrusion detection systems, when evaluating an interior system consider:

- Probability of detection
- Vulnerability to defeat
- Nuisance alarm rate

When deciding on active or passive sensors, consider the environment they are protecting. Interior active sensors have been known to sense a disturbance when the air conditioner or heater activates a stream of air in the protected area, especially if the air is causing curtains or drapes to move. Passive sensors on doors or windows may be defeated by a professional thief or extremist, but it will be more difficult to defeat active sensors inside of the structure, especially if they are covert. There are three application classes for interior sensors:

- Boundary-penetration sensors
- Interior motion sensors
- Proximity sensors

Boundary-penetration sensors detect penetration through the boundary of the protected area. These include vibration, electromechanical, infrasonic and passive sonic, capacitance, and active infrared sensors. They are used on doors, windows, other openings, and on walls.

Vibration sensors detect movement of the surface to which they are attached. They provide an early warning of a forced entry. *Glass-break sensors* attached directly to the glass are vibration sensors. Some of these are very simple. Others including fiber-optic sensors are more sophisticated.

Electromechanical sensors are passive, visible, line sensors. They basically consist of a switch unit and a magnetic unit and are used mostly on doors and windows. The switch unit is mounted on the stationary part of the door or window, and the magnetic unit is mounted on the moveable part. When the door or window is closed, a magnetic field is established between the units. When the door or window is opened, the field is broken and the sensor is activated.

Infrasonic sensors sense pressure changes. For example, when a door is opened (or closed) there is a light pressure change in the room. Since normal heating or air conditioning systems cause pressure changes, these sensors are best used in areas where there is limited access and they are unaffected by heating and air systems.

Capacitance sensors are usually proximity sensors although they can also be used for boundary-penetration detection. They are active sensors placed between a protected metal object and a ground plane, creating a part of the capacitance of a circuit in an oscillator. The object to be protected is electronically isolated from the ground. The sensor is activated when a person approaches the metal object. How close the person gets to the protected asset before the sensor is activated can be determined by the sensitivity of the control unit.

Active infrared sensors establish a beam of infrared light from a transmitter to a photo detector receiver. Several beams are usually used to establish a system of multiple beams. When the beam is disrupted, the sensor is activated. These systems are prone to nuisance alarms unless used in appropriate areas free of smoke and dust, moving objects such as animals, or curtains moving from air patterns.

Interior motion sensors include microwave sensors, sonic and ultrasonic sensors, passive infrared sensors, and video motion detection. These sensors create zones, and when motion occurs within the zone, the sensor is activated.

Interior microwave sensors operate on the same principle as exterior microwave sensors. The unit includes a transmitter and receiver. When the signal received by the receiver is different from that sent by the transmitter, the alarm is activated. Due to the technology involved, optimum detection occurs when the subject moves towards or away from the sensor rather than across the detection zone. A number of environmental considerations can affect the operation of these sensors, including fluorescent lighting. Microwave sensors should not be used in areas with fluorescent lighting unless filters are used that ignore the energy shift the lighting creates.

Active sonic sensors establish a detection field using a clearly audible tone. These tones are usually within the hearing range of the human ear and unpleasant. They are not normally used in facilities where a number of people are present.

Like microwave sensors, *ultrasonic sensor detection* is based on a difference between the energy transmitted and the energy received. A major feature of this sensor is that it will not penetrate walls or other physical barriers so that it can be contained within an enclosed area. Further, the wall may reflect the energy and enhance the detection within the zone. Again, air turbulence and acoustic energy sources may cause nuisance alarms.

Passive infrared sensors respond to changes in energy from a person entering the protected zone by detecting changes in thermal energy. These passive sensors have well-defined detection zones. They are usually low-cost (as compared to other systems) and, if properly installed, cause few nuisance alarms. However, the sensitivity will change as room temperature changes, and, since they are line-of-sight, the field can be easily blocked.

A *video motion detector* is a passive sensor that processes a video signal from a surveillance camera. Digital video motion detection systems divide the zone into numerous zones, elements, and cells. A change in any of these areas can trigger the alarm. Avoid analog systems on the market—stick with digital systems unless you want more equipment to add to your 8-track music and Beta video player collection.

There are two types of proximity sensors—capacitance proximity sensors and pressure sensors. Capacitance proximity sensors are active, covert line sensors; pressure sensors are passive and covert.

Capacitance proximity sensors operate on the same principle as electric capacitors. A change in the electrical charge or the dielectric medium causes a change in the capacitance. The capacitance proximity sensor monitors the capacitance between the metal object to be protected and a ground. When the capacitance changes, the alarm is triggered. The capacitance will change as the intruder approaches the metal object.

Pressure sensors can be placed around or underneath mats or other flooring. When an intruder steps on the sensor, the alarm is triggered.

Since a professional criminal or extremist is likely to step over the mat, it usually detects the more amateur criminal.

When considering an intrusion detection system, remember the three important questions to consider:

1. What is the probability of detection?
2. What are the system vulnerabilities?
3. What is the nuisance alarm rate?

Alarm systems

The sensor detects a possible intrusion into a protected area. Now what? Hopefully, an alarm is activated. The alarm tells you that an event may have occurred. More specifically, it should tell you:

- Where the alarm occurred
- Who or what may have caused the alarm
- When the alarm occurred

When planning an alarm system, you have to decide if the *enunciator* (*i.e.*, a loud noise) will be connected to the system. When the sensor detects an intrusion and the alarm is activated, do you want a siren or other noise to occur, or do you want the alarm to activate an alert at a monitoring center—or both?

Another decision is where the alarm is to be activated. Large utilities may have a proprietary security force who monitors and responds to alarms. Other companies will use contract security companies who monitor the system and notify local police and the utility when an event is triggered.

One of the components of an alarm is its communications system. The signal must travel from the sensor to the alarm to allow a response to the detection. Some of these systems use radio communications while others use wire or fiber optics. The communication's system must be secure and redundant.

The components of the alarm system can be connected using different network configurations. These include:

- Point-to-point
- Star network
- Loop network
- Bus wiring configuration
- Ring configuration

The simplest is the *point-to-point,* in which devices (including sensors and alarms) are connected directly to each other. This is the easiest configuration to use.

The *star configuration* uses a collection of point-to-point connections to permit sensor data to be communicated to a field panel. The star configuration is easy to understand and to use, but it is not redundant. However, because the communication between the individual sensors and the alarm system is independent, a single point failure only disables one part of the system. The disadvantage to the star configuration is the amount of cable required for the installation and potential problems with adding additional sensors if there is no connection on the receiver for additional cables.

Loops use point-to-point connections in an interconnecting loop. They start and end at the same physical location. Bi-directional loops provided redundancy and are more efficient than star networks.

In a *bus network,* devices share the same common media. While bus networks are efficient, they are not as reliable as other configurations, because a single device failure can result in a system failure.

The *ring configuration* combines the loop and bus approach. Devices are connected separately to the loop, but because the communication is through the loop, redundancy and reliability increases.

Regardless of the configuration, you should have the capability of testing the communications between the sensors and the reporting system. The system should indicate when there is a problem without a query, but you should also have query capability.

If you have an in-house monitor and alarm response system, then consider the use of surveillance cameras in the areas where the sensors are installed.

Surveillance cameras

When considering a closed circuit television (CCTV) system, you should evaluate the sensitivity of the light available in the zone and the ability to maintain an adequate picture as the light changes. The components to a CCTV system include:

- Camera and lens
- Lighting system
- Transmission system
- Switching equipment
- Monitor and recorder
- Control system

CCTV deters crime. When the school system in a major U.S. city installed cameras at each of its elementary and secondary campuses, burglaries were reduced by 50% in one year. The savings in insurance premiums help to offset the cost of a system. However, CCTV has also resulted in bargaining grievances and legal actions against employers. In many cases, courts and regulatory agencies have ruled that cameras in the workplace do not infringe upon a reasonable expectation of privacy. A decision by the National Labor Relations Board in April 1997 ruled in a case involving Colgate-Palmolive that the installation of video cameras should be considered a mandatory subject of bargaining. In this case, the ruling involved the use of hidden cameras.

So, before hidden cameras are used for internal investigations, discuss their use with your legal and security departments; even then, they should be used only if probable cause for the investigation has been established. Since the objective of the use of CCTV in most utilities is to deter crime, the cameras should be clearly visible. There should also be signs on the building at all entrances stating that the facility is under video surveillance 24-hours a day.

Three basic rules govern use of a video surveillance system. First, keep it simple. Second, don't buy yesterday's technology. Finally, decide in advance where the monitor and recorders will be located.

There are systems that pan, tilt, zoom, and allow for control from the control panel. If you have a full-time guard force monitoring critical zones, these may be necessary. In most cases, you need a simple application with a fixed camera with the appropriate lens. Consider also whether you really need color cameras or if black and white will serve your application. The more bells and whistles on the system, the more that can go wrong.

There are some good buys on analog CCTV systems, but that's because no one is buying them. You should only consider state-of-the-art digital systems. This is a field where technology is constantly changing. Digital systems have different media recording capabilities, and there is an evolution occurring in image sensors. Do your research and then ask for suggestions and bids from several professional CCTV companies.

Your system should have a recording capability. Digital systems can record for long periods of time and require little maintenance. Imagine changing the videotapes on an analog system every 24-hours! Frankly, if you're going to install a CCTV security system, it doesn't make sense to leave out recording capability. Sometimes you will not know an incident has occurred until hours later. The video recording will provide you with information regarding the event that should lead to the correct conclusion.

So, where do we put the monitor? In one utility, it sits in the human resource director's office behind a stack of HR manuals. In another utility, it was in the customer service manager's office, behind the desk where this person couldn't see it. The monitor should be positioned where someone is actually *looking* at it, even if just out of the corner of his or her eye. This may seem like an obvious consideration, but it is an important one.

Summary

Physical security and access control can be as simple as locking the locks you already have on entrances and exits, or it can include sophisticated intrusion detection and CCTV systems. Look at your facility building plans and identify all entry points. Start from the outside and work in using a concentric circle approach. Begin with clear areas outside the fence, the fence itself, clear areas inside the fence, the building entrances, and access control within the building.

Your objective is to delay the adversary. The time required to reach key assets must be long enough to allow police or the guard force to intervene. Where can you add additional barriers or controls?

Your system should be integrated. If possible, use the same vendor for your access-control systems, intrusion detection systems, alarms and response, and CCTV. These systems must work together if your security objectives are to be achieved.

Finally, remember that you need to ensure that *intention equals reality*. There should be an annual inspection of these systems and unannounced checks. Reports on vulnerabilities and needs should be carefully documented and submitted to management. Don't wait for incidents to occur. To be effective, your physical security and access-control program must be proactive.

Action Checklist

1. How much physical security you need depends upon the threat.
   a. For normal threat conditions consider the following:
      1) Protective barriers including clear areas, fencing, perimeter entrances, and exits.
      2) Protective lighting.
      3) Structural security including building construction, doors, and windows.
      4) Access controls.
   b. If there is a terrorist or extremist threat, you should also consider:
      1) Removing labels on buildings and parking spaces.
      2) Adding barriers to high-threat areas.
      3) Upgrading the intrusion detection in high-threat areas.
      4) Manning critical sites during increased levels of threat.
2. You can't protect everything all of the time.
   a. Look at the building plot or fire evacuation plan and identify physical security and access controls already in place.
   b. Identify the critical areas or offices in the building.
   c. Recommend measures to increase the security at the critical areas and offices.
   d. Be prepared to increase the security at critical areas during increased levels of risk.
3. Physical security standards.
   a. Use Table 7–1 and Appendix F to evaluate the physical security at this facility.
   b. Visit the facility at night to assess the protective lighting. Use Table 7–2 to assist in this evaluation.
   c. Determine the vulnerability of the structural security using Table 7–3.

4. Assess the utility of the "bells and whistles" currently in use.
   a. Intrusion detection system.
      1) External intrusion detection systems.
      2) Internal intrusion detection systems.
   b. Alarm system.
   c. Surveillance cameras (CCTV).

Notes

1. Garcia, Mary Lynn, *The Design and Evaluation of Physical Protection Systems* (Woburn, MA: Butterworth-Heinmann), 2001

8

Protecting Employees and Physical Assets

Basic Protection Considerations

Employees are the most important assets the utility has. In this chapter, we will discuss how to protect employees as well as physical assets. Unfortunately, we may have to protect physical assets from the employees. While most people who work for utilities are outstanding workers, a few decide to steal company property, embezzle company funds, or engage in vandalism for some perceived injustice and must be dealt with.

In chapter 6 we discussed access control. We begin this chapter with an extension of that discussion—entry control. Although entry control is a component of access control, we will discuss it here as a method of protecting employees and physical assets.

According to Garcia, the objectives of entry control are:[1]

- To permit only authorized persons to enter and exit
- To detect and prevent the entry or exit of contraband material (weapons, explosives, unauthorized tools, or critical assets)
- To provide information to security personnel to facilitate assessment and response

The level of sophistication needed in your entry control system depends upon the threat. Many utilities use a simple keypad system; punch in the right numbers, and the door opens. Others use employee identification or proximity cards. The employee identification card must be swiped into the card reader; the proximity card needs only to be held near the reader. Where high-level secure entry is required, entry procedures may combine employee identification or proximity cards and a personal identification number.

There are benefits and problems with each of these systems. Access via an employee identification or proximity card allows anyone who has access to the card to gain entry. In one case, a group of government office workers arrived at work on Monday morning to find that none of their 70 computers worked. On Saturday, someone had entered the office and removed the memory chips and hard drives from every computer. The office was located on an upper level of an office building that required the use of a personal identification card to gain access on the weekends. The computer identified a worker in another office on the same floor who had entered the building for an extended period of time on Saturday. This worker claimed that she had not been in the building on Saturday. When the police went to her home, they found her teenage son with the stolen merchandise. He had removed the card from his mother's purse without her knowledge and replaced it after stealing the computer components.

Employees will resist an entry control system that is inconvenient. Do you really need keypad access, a proximity card, and a personal identification number to gain access to a building? If there are critical assets or controls in the building, the answer may be yes. Otherwise, it may be overkill. For most utility entry systems, a simple but effective approach is best.

The biggest problem with most entry control systems is *piggybacking*. You have established an employee identification card system that records the time and date of each employee's entrance and egress to the facility; but, when a number of employees arrive at work at the same time, one uses his/her card to open the door, and the rest *piggyback* inside without using their badges. This may be convenient for the employees, but it usurps the purpose of the system as much as propping open secure doors. Implement the three-step program below to manage this problem.

1. Sell employees on the need for security and the value that these systems have in protecting them.
2. Monitor use of the system. Watch employees as they arrive or leave the facility, and identify employees who bypass security systems and procedures.
3. Discipline employees who bypass security systems and procedures. They are putting other employees, customers, and assets at risk.

Biometrics or *Personal Identification Verification systems* are the emerging technology in entry control-systems that confirm the identity of persons attempting entry to a secure area by corroborating a known personal characteristic such as fingerprints or hand geometry. Other systems use handwriting analysis for confirmation, usually the individual's signature. Sophisticated systems confirm voiceprints and eye patterns. As this technology improves and becomes less costly, it may be used in utility high-security areas, including computer centers and cash-handling areas.

Should you include security guards in your security program? Again, the answer to this question depends upon the level of threat. If you have a lot of cash in a customer service area, then guards may be required. If you have valuable assets at a warehouse, then there maybe a need for guards there as well.

Ask two primary questions when considering the use of security guards. First, should you have a proprietary (*i.e.*, employee) guard force or use contract guards? Second, should these guards be armed?

Maintaining an effective guard force requires careful personnel selection, training, and performance monitoring. Since most guards are paid only $8 to $12 an hour, this becomes a challenge, but professional guard companies have the experience and administration required to meet these challenges.[2] If you contract with a guard company, ask about their selection, training, and monitoring programs. Also, be sure they have appropriate bonding and insurance. Unless you are a large organization with a significant security department, contract guards are preferable.

Arming guards presents additional concerns. Armed personnel have to be licensed and trained. They are required to requalify with their weapons at least once a year. They also introduce additional liability. If they use their weapons while protecting utility personnel or property, expect to spend a great deal of time with your lawyers. If the threat is such that you feel the need for armed guards, contact your local law enforcement agency and consider using off-duty police officers. If you need armed personnel on a continuous basis, some of the professional guard companies also have qualified personnel. Many of these people are former police officers.

All of these elements—physical security and access controls, entry control, and guards—can only be effective if employees commit to the security program and assume personal responsibility in contributing to its success. For your program to be successful you should:

- Include a security briefing during new employee orientation programs
- Provide an annual security briefing for all employees
- Openly monitor the security program to identify vulnerabilities and problems
- Encourage employees to provide suggestions and input into the program
- Discipline employees who violate the program's polices and procedures

Protecting Employees

Inside employees

For purposes of this discussion, *inside employees* are those whose majority responsibilities are carried out in an office or warehouse environment. *Outside employees* are those who work primarily in the field. The challenges in protecting these two separate categories of employees are different. For example, while employees in either category may be confronted by an angry customer, the inside employee usually has other employees present, whereas the outside employee is on the customer's turf and usually alone.

When dealing with an angry customer in the office, get the customer out of the lobby area and into an office. Have them sit down, and give them the opportunity to share what's on their mind. They may have a legitimate grievance that needs to be addressed, or they may be displacing anger resulting from a totally unrelated matter. In either case, they need to vent some of the anger before they become rational again. To summarize, when dealing with angry customers in the office:

- Get them out of the lobby and into an office
- Get them to sit, preferably on the opposite side of a desk or table from you
- Let them talk (the venting process)
- Listen (if you have been trained in active listening, this is the time to use those skills)
- Try to reach some resolution with the customer

Any utility employee who has worked in customer service for any period of time should have developed these skills. If not, they're in the wrong position. A final suggestion: the employee dealing with the angry customer should have some way of notifying others if the situation gets out of control. Using glass instead of solid walls in customer service offices helps to accomplish this. *Panic buttons* under the desk are another approach. The panic button would alert another employee who then enters the office. Rather than asking, "Is anything wrong here?" The responding employee should ask, "Can I be of help?" The statement should not suggest that something is wrong, but rather that the responding employee is there to assist in resolving the situation.

If threats are made during the discussion, these should be recorded. Most threats made in anger are forgotten, but, on occasion, the customer acts on them. Don't challenge the threat. Record it, and discuss it with management later. At that time, a decision must be made as to the seriousness of the threat and the appropriate response.

Inside employees often deal with angry customers on the telephone. Again, let the customers vent, and listen to what they are saying. In most utilities, if the customer becomes abusive on the telephone, the policy is to suggest they call later when they have calmed down, and to hang up. If threats are made, these should be recorded and discussed with the supervisor.

Among the best deterrents to armed robbery are the use of surveillance cameras and an alarm system. (But if surveillance cameras were an absolute deterrent, criminals wouldn't rob banks.) Robbers target businesses where there are poor cash-handling procedures, poor housekeeping, and a failure to plan on the possibility of being targeted for a robbery. Most robberies take place in less than one minute. During that time, employees need to comply with the robber's demands, remember as much as possible about the robber(s), and be alert to threats of physical harm.

There are a number of actions you can take to decrease the probability of a robbery at locations where cash in handled:

- Keep the exterior of the building well lighted.
- Keep signs out of the exterior windows so that the interior is seen from the outside.
- Keep unauthorized access doors to the area locked.
- Ensure that the alarm is working.
- Keep cash exposure and cash on hand at a minimum at all times.
- Make sure that customers at the inside counters cannot see the cash drawer at the drive-in window and that customers at the drive-in window cannot see the cash drawers of the internal cashiers (a common vulnerability at many utilities).
- Keep checks separate from cash when making bank deposits.

If you are not using a courier service (*i.e.*, armored car service) when making bank deposits, go directly to the bank, and conceal the deposit. There is nothing more inviting than seeing the same employee leaving the utility at the same time every day with a deposit bag as they drive, and sometimes walk, to the bank. This really happens! Don't be predictable.

If possible, don't make the trip to the bank alone. Many utilities have established a procedure by which a police car picks up the employee with the deposit and drives the employee to the bank. This policy is often used in municipal utilities since the police department and the utility are both a part of city government.

Even with all the robbery prevention procedures you put into place, you may still be targeted. If a robbery does occur, employees should comply with all of the robber's demands. If the robber displays a firearm, assume it is loaded. If they do not display a firearm, then assume they have one. If possible, activate a silent alarm that alerts your alarm company or the police.

Common sense says to do everything robbers tell you to do, but there may be a decision-making time. Many robberies are committed to support the criminal's drug habit. These criminals are less predictable and potentially more violent. Two examples:

The driver at a motel left at about 4:00 A.M. to pick up a train crew. When they returned, two drug-crazed robbers had the motel clerk on the floor behind the counter and were apparently about to kill him. Startled by the arrival of the driver and the four members of the train crew, the criminal with the gun came up from behind the counter and told everyone to freeze. They all did as they were told. He then decided there were too many witnesses and began shooting, starting with the train crewmember to his left. The person to the extreme right realized the gunman was going to kill them all and ran down a hall, out of the motel, and into the woods. The criminals panicked and ran out the door without killing any of the other witnesses. The shooter was arrested some months later in a shooting gallery (a place to buy drugs and use them). He is currently serving 35 years to life. His accomplice was never apprehended.

The second incident took place in a restaurant located in a high crime area. Security procedures at the restaurant required that the rear door was never opened unless the employees knew who was requesting access. The security also included CCTV in this area since it was where the cash was counted. One morning, three employees were preparing for opening when there was a bang on the back door. Ignoring security policy, one of the employees opened the door, and two masked robbers came in. They took the opening cash and then told the employees to go into the storeroom where they would be tied up. All of this was recorded on the security monitor. What the monitor also recorded were the three shots that were fired when the employees were executed. The robbers dropped the money while running out the door. They were never caught.

The point is that in today's world, the response to a robber may not always be to "do exactly as told." If the robber tells you we're going in the back room or everyone needs to get down on the floor, it may be decision-making time. If your life is in danger, you may need to decide to run, to try to tackle the robber, or to take some other action to save your life. These situations are the exception, but they do occur. If you are the victim of a robbery, only you can decide if lives are at risk.

In general, your response to a robbery should be:

- Do not take any action that would risk your life or the lives of others.
- Do exactly what the robber(s) tells you to do.
- If possible, activate a silent alarm.
- Attempt to alert other employees to the situation using prearranged signals.
- Remember as much as you can about the criminal.

If there are several robbers, do not try to remember as much as possible about each of them. Pick one and focus on that person. Use the nickname approach. Who does this criminal remind you of? Uncle Louie with the unusual nose? Mickey Mouse because of his large ears? Remember the nickname and why you selected it. When interviewed by the police, it will help you to recall details regarding the robber's description.

Some of the actions you can take to prepare for a robbery are listed in Table 8–1. Some of the actions you should take during and immediately after a robbery are listed in Table 8–2.

After the robbery, immediately lock down the area, and call the police. *Do not* discuss the incident with fellow victims. Each person present should put his/her version of the events and the description of the criminal(s) in writing. The police will want this information as they interview you individually. When you get home and the media calls you to ask about the incident, tell them to call the police department. *Do not* discuss the events with the media or with friends. You may need to discuss it with family members or a counselor to deal with the trauma of the event. This is normal. Don't try to deal with the trauma of the event on your own. Many police jurisdictions have victim's assistance personnel trained to help you. Ask for them.

Table 8–1 Actions to Consider Before a Robbery

ACTION	COMMENT/ASSIGNMENT
Be alert for suspicious persons in or near the office at all times.	
Instruct all employees in the use of the alarm system.	
If using bait money, instruct all employees in its use.	
If there is a suspicious vehicle in the area, note the license plate number and description, and call the police.	
MAKE PLANS IN ADVANCE AS TO WHO WILL BE RESPONSIBLE FOR THE FOLLOWING ACTIONS:	
Call the police.	
Protect the evidence at the scene.	
Detain customers and other potential witnesses.	
Notify management.	
ADDITIONAL CONSIDERATIONS	
Practice identification with employees.	
Install height markers, e.g., lines at various 6" intervals, on the door frame ranging from 5'0" to 7'0".	
Discuss with employees what they might do if a robbery occurs.	

Table 8–2 Actions to Take if a Robbery Occurs

ACTION	COMMENT/ASSIGNMENT
OBSERVE THE PHYSICAL CHARACTERISTICS OF THE ROBBER(S)	
Race, age, and height.	
Facial characteristics, complexion, and hair.	
Clothing worn.	
How did the person carry him/herself?	
Speech.	
Scars, marks, or deformities.	
Method of operation.	
OTHER CONSIDERATIONS	
Look for accomplices.	
Note the direction and method of escape.	
Describe the type of weapon used.	
Use the "nickname" method to remember the characteristics of the robber nearest to you.	

Occasionally, an armed robbery attempt becomes a hostage situation when the police arrive while the criminals are still in the building. An angry customer or domestic dispute can also lead to a hostage situation. Most police departments have personnel trained to deal with these situations, and most hostage negotiations are successfully resolved with the release of the hostages and the arrest of the criminal.

Hostage negotiations are most successful when the situation is contained to one area of a building. Since this type of incident is most likely to occur in the cashier or lobby area, all of the doors from these areas that lead into rest of the building should have entry controls, such as keypads. If you use employee or proximity cards, then the criminal will simply take a card from one of the hostages and gain access to other areas. Your objective in planning security for these areas is to be prepared to isolate and contain any situation. There are several other steps the utility should take to prepare for a hostage situation:

- ***Have a complete floor plan of the area for the police.*** The floor plan should show the details of the area including counters, desks, and other furniture. It must show all of the windows, doors, and other openings and the direction that the doors open. It is helpful if the doors are labeled, *e.g.*, A, B, C, etc., and it is also helpful if labels are clearly visible on the outside of the building.
- ***Do not communicate with the hostage-takers while waiting for the police.*** If they attempt to telephone your office or communicate using any other media, stall. Don't make any promises or commitments, and don't ask what their demands are.
- ***Provide complete information on employees and others being held hostage.***
- ***Provide information on the hostage-takers and how the situation started.***
- ***Dispatch a manager or supervisor to the family of each hostage employee.*** The police will send an officer to question the families. The manager or supervisor's role is to provide support.
- ***When the situation is resolved the police will treat the area as a crime scene.*** Have an alternative location to conduct the activities normally conducted in this area.
- ***Be prepared to support the employees and other persons held hostage and their families.*** At the very least they should have access to initial counseling. In some cases. they may require additional counseling to deal with the trauma of the event. There are professional counselors who specialize in this area.

Employees held hostage should do what they are told and not become confrontational with the hostage-takers. Don't stare at them or use nonverbal communication that may appear to be confrontational. When the situation is resolved and they are released, the police will verify their identities and then question them. There have been incidents where the hostage-takers have tried to walk out pretending to be hostages as part of the escape plan.

In those rare cases where the situation becomes violent or potentially violent, hostages should get down on the floor behind something. Stay there until the police tell you to move. The police may put restraints on the hostages until they can verify their identifications. This is normal procedure.

Outside employees

As stated, outside employees are at a greater risk of attack from angry customers because they are on the customer's turf. The customer has the *home advantage* and may feel that he/she is protecting his/her property. The other danger is that the customer may have a weapon in the house and decide to get it.

Always try to avoid being confrontational with customers in the field. If the confrontation appears to be unavoidable, try to leave as quickly as possible. When arriving at a customer location (especially at a service where the account is being disconnected for nonpayment or some other activity that could result in a difficult situation with the customer), always have an escape route planned before approaching the service. Thinking about escape routes should become routine for employees who perform these duties. Back your vehicle into the parking space or driveway, and also look for escape routes by foot. Again, this should be routine for utility employees in certain positions.

Some utilities still collect money in the field. As discussed in an earlier chapter, this is not a good idea in today's world. Don't collect checks in the field either. If crooks see your employee accepting checks, they may assume the employee collects cash as well. Even if they don't make this assumption, they may want the checks for a *check-washing operation* (a criminal enterprise where ink is removed from the check, a new amount and payee entered, and the checks are cashed). This is big business in some areas of the United States. A number of utilities have experienced break-ins at their post office boxes, usually on weekends, by check-washers.

Employees who disconnect services for nonpayment or conduct theft-of-service investigations are obviously at risk for having to deal with angry customers. But *all* field employees should be prepared to deal with these situations.

Meter readers, service people, and other employees occasionally find marijuana patches or other drug operations. Local police should be invited to conduct a training session for all outside employees once a year to show them what to look for at "growing operations" and drug laboratories and how to respond if they see evidence at a service location. (The most immediate response is to leave.) Utility employees should not return to a location where there has been a potentially violent confrontation or a crime committed unless accompanied by law enforcement. The police go in first, and the employee follows once the scene is secured.

Some utilities maintain a *danger account* list. These are services where criminal activity is suspected, where customers have made threats to utility employees, or where the customer is extremely difficult. This includes customers who chain vicious dogs to the meter or who erect fences that prohibit the utility from gaining access to the meter or other parts of the service connection. It also includes customers who are stealing utility services. Several suggestions for dealing with dangerous accounts:

- The fact that the account is designated as dangerous and the reason for making this decision should be a part of the customer's record.
- Any work order or other assignment that sends an employee to the location should indicate that it is a dangerous account.
- Employees should not go to the service alone. There should always be two employees.
- If there is a safety concern, employees tell the dispatcher they are going to the location and provide a specific time when they will call to say they have left. If the second call is not received, the police should be notified.

Dealing with angry people

Utility employees need to know how to deal with angry people of all sorts. In most cases, the angry person is a customer, but in some cases, it may be a fellow employee. It is important to understand that anger is not just an emotion—it is a physical response. When a person becomes angry, certain chemicals are released into the bloodstream activating the nervous system. The angrier the person becomes, the more chemicals are released and the longer it takes for the person to "vent" as these chemicals are dissipated and the nervous system returns to baseline. That is why saying to someone who is extremely angry, "Now, calm down and let's discuss this rationally," is not going to work. The venting process allows the chemicals time to dissipate. The worst possible response to anger is anger. If an angry person yells at you or threatens you, and you respond by yelling or threatening, what happens to the amount of chemicals released into the body? It elevates, and the person becomes angrier. The escalation could lead to a physical confrontation.

To defuse an angry person and avoid a possible physical confrontation:

- ***Remain calm, at least outwardly.*** Be aware of your body language and avoid assuming an aggressive or threatening posture. If you are standing, put one foot in front of the other, not parallel to each other. This is a better position from which to defend yourself if the angry person does become aggressive. It's also a better position from which to run if that becomes necessary. If the customer begins to become aggressive, put your hands at about waist height, palms out. This is a defensive position that tells the angry person you are not responding to their aggression with aggression. It also defines your "personal space."
- ***Go into your active listening mode and encourage the person to vent their angry feelings.*** Listen for the real reason for the anger. In one case, a customer was aggressively complaining about a bill, saying that he was home all day and never saw a meter reader at the house. As he was venting, he also said that he was out of work and had recently suffered weather-related damage at his house. Because the customer

service representative allowed the customer to vent and listened to what he was saying, she was able to identify the real problem, arrange a payment plan, and show him ways of decreasing energy usage.

- ***If possible, get the customer seated.*** People in a seated position are less likely to become physically violent.

Some people—especially those who are criminals or have certain mental illnesses—have confrontation-prone personalities. These people are different from a normal person who becomes angry, and the potential for violence is greater. Individuals with an *inadequate personality disorder* may be confrontation-prone. In some cases, this is passive-aggressive and less likely to lead to physical violence; in other cases, the person may be emotionally unstable. Most of the people in this category know that the aggression and violence is wrong, but they commit the acts anyway.

Individuals with *antisocial personality disorders* may not understand that their aggression or violent acts are improper. It releases tension they may experience, and it often gets them what they want. Many criminals fall into this category. A nonviolent antisocial personality may be a career burglar who sees nothing wrong with breaking into your house and stealing things that you have worked hard to obtain. In their mind, it is how they earn their living. Individuals with a violent-prone antisocial personality disorder may physically assault you, shoot you, or otherwise attack and afterwards suffer no remorse. These people lack a conscience. They can be dangerous.

Certain mental disorders (including bi-polar disorders and paranoid schizophrenia) can cause a person to become violent. *Bi-polar* used to be referred to as *manic-depressive*. Individuals with this illness suffer mood swings. At the depressive end, the individuals are potentially dangerous to themselves. At the manic end, they may be a danger to others if provoked. Paranoid schizophrenics are potentially dangerous if they perceive you as a threat. Fortunately, medication and treatment for these disorders is available; but if the person is not taking the medication, he/she may present a threat.

If you suspect that the person you are dealing with has a confrontation-prone personality, take the following actions:

- Remain outwardly calm. Be a role model.
- Do not believe anything they tell you.
- Disengage (leave) as soon as possible. If you are in the office, make sure other employees are present. If you are in the field, try to leave the scene. Tell the customer you need to check on something back at the office.

Protecting employees from themselves

A final consideration in protecting employees is that we sometimes need to protect them from themselves. A utility's warehouse was filled with enough conductor to last for almost 10 years. When the purchasing agent was asked about it, he said he got a good buy. Then he and his wife went to Hawaii for a week's vacation.

A manager (male and married) in a utility had an affair with a customer service representative (female, not married). When she applied for the supervisor position in customer service and didn't get it, she filed a sexual harassment complaint. Was there to be a quid pro quo? The case was settled out of court, and she can now afford to go to Hawaii on her own.

Six employees were fired at a municipal utility on allegations of "inappropriate sexual conduct on town property or town time, racial derogatory language, and theft or misuse of town property." Only one employee appealed the action. Witnesses said they saw him having sex with another town employee in his car during working hours, but he claims the car was in the shop that day.

A utility employee had a company-owned video camera to be used to assess job training and safety, even though he was not in that department. He was arrested early one morning videotaping through a woman's window, as she was getting ready for work. When the police looked at the tape, there were several other women who had been videotaped through their windows.

The last incident was obviously illegal. But were the others? If they were not illegal, they were certainly unethical. Many companies, including utilities, require employees to attend an annual ethics briefing during which basic do's and don'ts are discussed. At the end of the briefing, employees sign a statement certifying that they attended the briefing and understand the ethics requirements discussed. If they are found to violate these ethical standards, they are subject to disciplinary action. The ethics statement should be prepared by the utility's attorney.

Protecting Physical Assets

A father was surprised to receive a telephone call from his son's school principal. The son had been caught stealing pencils, paper, notebooks, and computer supplies from the school bookstore. "Why would he do that?" the father replied. "I bring all the office and computer supplies he needs home from work!"

Retailers admit that they lose 10 times as much from employee theft than they lose to shoplifting. But do utilities have an employee theft problem?

- A town official in charge of utilities was forced to resign when he was charged with utility theft and burglary. He is alleged to have rigged the electric meters at seven properties he owns and stolen $75,000 worth of electricity and gas in a 10-year period. The service at these locations is provided by another company, and he paid an employee of that company $50 to ignore the theft.
- The head of a small utility in the Midwest admitted that he charged between $4,000 and $5,000 in personal telephone calls, meals, and hotel accommodations to his utility in a single year. He pretended to be at utility-related meetings out of town but was actually with a family friend.
- At a utility in the Northeast, a similar situation: a utility manager is accused of living lavishly with two other managers on a trip to Las Vegas. They stayed in the best hotel, rented a Lincoln Town Car,

and ate meals and drank wine that Donald Trump would have envied. Allegedly this was not an isolated incident.

- Employee theft at utilities is an equal opportunity crime. A utility clerk who was eight months pregnant was charged with stealing $37,000 from the utility. She was arrested at the home of another city employee who is being charged with stealing from the city.
- Employee theft is not limited to municipal utilities. A 20-year employee at an investor-owned utility was charged with embezzling $776,212 during an 18-month period. She needed the money to pay for her new boyfriend's cars, clothes, and other expenses. He lived well. She went to jail.
- Not all employee thefts involve money. Five employees at a large utility were ordered by a court to return documents they had stolen before they left. They used this proprietary information to start their own company. Their former employer is also suing them for damages.
- An employee used the utility's truck to steal patio furniture from behind a business. The owner saw him load up and drive off and followed him in his car while calling police on his cell phone. The employee tried to lose the storeowner but was stopped by police.
- It is most embarrassing when the person stealing from the utility, especially a large investor-owned utility, is the chairman. The investigation of the larceny, bribery, and illegal campaign contributions charges cost this company at least $6 million. The chairman was not alone. A vice president at the company (who was also a former state senator) was charged with threatening to pull a million-dollar contract from the utility's advertising agency unless the president of that company made a $25,000 contribution each year to certain political parties. Two other former employees of the utility, a husband and wife, were arrested for grand larceny. They had a printer submit bills for $13,912 and $18,761 for brochures that were never printed. The printing company was paid, and the money then given to the two former employees.

According to the Association of Certified Fraud Examiners, the costliest abuses take place in companies with fewer than 100 employees. They estimate that fraud and other employee crimes cost employers in the United States more than $400 *billion* per year, or an average of $9 per day per worker. Men, who account for 54% of the workforce, are responsible for 75% of the abuses.

Some employees have a history of stealing from their employers. These people will hopefully be identified during the hiring process and not offered a position. There are a number of cases, however, where long-time trusted employees engage in theft or fraud. The illegal behavior may result from a change in lifestyle (like the 20-year employee who stole $776,212 to help her new boyfriend maintain his lifestyle), a personal financial crisis, a drug or gambling habit, revenge against a perceived injustice, or a nonfinancial personal crisis.

A number of things can be done to encourage honest employees to remain honest and to detect dishonest employees:

- ***Take a careful look at your hiring process.*** Let applicants know that you do background and credit history checks on all new employees. The credit history check is often more informative than the background check. Consider using a valid written integrity test.
- ***Include the need for honesty and integrity in new employee orientation programs.*** Make sure the new employees understand what behaviors are acceptable and what are not acceptable. Discuss the code of ethics.
- ***Install security systems and implement internal controls for any process involving money or other valuable assets.***
- ***Encourage positive management, supervisory, and human resource programs throughout the utility.*** These include employee development, communications, and performance evaluation programs.
- ***Identify and counsel employees with problems.***
- ***Discipline employees when they violate the rules.***
- ***Review compensation packages to ensure internal and external equity.***
- ***Make vacations mandatory.***
- ***If possible, rotate assignments.***
- ***Implement simple controls to monitor inventory, expense accounts, purchasing, and other vulnerable areas.***

- ***Stay in touch with your employees.*** Get to know them, their needs, and their motivations.
- ***Establish a culture of honesty.*** If the chairman of the company is stealing, why should we be surprised to find other employees engaged in illegal activities?

Preventing outsider theft

Burglars don't like alarm systems. Therefore, you should have alarm systems and signs around buildings warning criminals that you have alarms and video surveillance. Think about which doors really need to be unlocked and during what hours or activities. Otherwise, keep them locked.

Good housekeeping is essential to burglary prevention. When closing the building at night and activating the alarm system, turn on lights inside and outside of the building. Make sure that no one is hiding in the building before you leave, and double-check the locks on all windows and doors.

Since most business burglaries occur at night, there are a number of simple things you can do to encourage the burglar to go elsewhere:

- The area around the utility should be well lighted
- Illuminate all entry points
- Keep night lights on inside the building
- Place a night light over the safe
- Leave empty cash drawers open

Asset identification systems will help to reduce insider and outsider theft. Every asset, from tools to computers, should be easily identified as the property of the utility. There are a number of quality, easy-to-use systems available. Some of these are tags that permanently affix to company property. Others are highly sophisticated and will set off an alarm or other notification if the property is moved or taken out of the building.

Summary

Your employees are your most valuable assets. Because of the utility's unique position in the community, your employees are at risk from angry customers and criminals such as drug growers and utility thieves. Personal security should be included in every new employee orientation and updated during periodic workplace violence prevention briefings.

Appendix E includes personal security checklists. These can be used as appropriate for employee training.

Physical security, access control, and alarm and intrusion detection systems will protect your physical assets. However, your employees also present a threat to these assets. The utility must have good internal controls and an asset identification program to protect its physical assets.

Action Checklist

1. Evaluate your basic protection considerations.
   a. Entry control systems.
   b. Selling employees on the need for security.
   c. Evaluate the potential use of security guards during normal operations and during elevated levels of threat.
2. Protect inside employees.
   a. Have specific procedures for dealing with angry customers in the office. Get them out of the lobby area and into an office where they can vent without being seen by other customers.
   b. Consider the use of "alert" buttons under customer service desks and counters. These would signal a manager or security personnel that assistance is needed.
   c. Develop an armed robbery prevention program.
   d. Train employees to respond correctly if a robbery does occur. This includes post-incident procedures.
   e. Prepare for a hostage situation. Have a floor plan available for police, and have the ability to isolate the area where this situation is most likely to occur.
3. Protect outside employees.
   a. Train outside employees to deal with angry confrontational customers.
   b. If you are collecting payments in the field, stop this practice. You are putting employees at risk.
   c. Train field employees to be especially alert at potentially dangerous accounts. These include drug growers and dealers and theft-of-service.
   d. Develop and maintain a danger account list. Employees visiting these locations should take special security precautions.

4. Train all employees to deal with angry customers.
    a. Remember that anger is not just an emotion—it is a physical response.
    b. Employees need to be trained in active listening so they can allow the angry person to vent.
    c. Ensure that employees understand the difference between a normal person who becomes angry and a person with a confrontation-prone personality.
5. Develop an employee ethics program.
    a. Your attorney should develop an ethics statement of specific do's and don'ts.
    b. Employees should attend an annual ethics briefing and sign the statement prepared by the attorney.
6. Develop an employee anti-theft program.
    a. Review your hiring practices, and make adjustments as needed.
    b. Establish or upgrade internal controls as needed.
    c. Establish a culture of honesty.
7. Protect the utility from external theft.
    a. Review burglary prevention procedures.
    b. Initiate or review and consider upgrading your asset identification program.

Notes

1. Garcia, Mary Lynn, *The Design and Evaluation of Physical Protection Systems* (Woburn, MA: Butterworth-Heinmann), 2001

2. Weisul, Kimberly, "Up Front: Heartland Insecurity: On the Prowl for Guards," *Business Week*, November 12, 2001, p. 16

9

Special Threats

Bomb Threats

Most bomb threats are hoaxes. Some are real. Waiting until you receive a bomb threat before developing a response plan can be disastrous. You need to be prepared to make decisions regarding evacuation and searches. You also need to have an established liaison with the police and fire departments with an understanding of their role if a bomb threat is received.

Bombs can be constructed to look like anything and can be placed at or delivered to the utility in any number of ways. Some bombs and threats are made to the utility's offices; others, like those of the NWLF, may be placed near transmission towers or other remote locations.

As with other types of threats, the utility should have a threat planning team and a response plan. The Bureau of Alcohol, Tobacco, and Firearms (BATF) recommends that you consider the following when preparing your bomb threat response plan:

1. ***Designate a chain of command.*** Use the same chain of command you use for other emergencies.
2. ***Establish a command center.*** This is usually where your communications center is located. You will need an alternate center in the event that the primary is the target of the threat or incident.
3. ***Decide what primary and alternate communications will be used.*** These include internal communications and communications with police and fire departments.
4. ***Establish clearly how and by whom a bomb threat will be evaluated.*** Not all threats will result in a complete evacuation. Some will result in partial evacuations and others in no evacuations at all.
5. ***Decide what procedures will be followed when a bomb threat is received or device discovered.***
6. ***Determine to what extent the bomb squad will assist and at what point the squad will respond.*** Many bomb squads do not respond unless a suspect device is found. The initial officers responding will be patrol officers. Unless a device is found, utility personnel will be responsible for deciding how to respond to the incident and for conducting the search.
7. ***Provide an evacuation plan with enough flexibility to avoid a suspected danger area.*** If possible, avoid evacuating to a parking area. The bomber may have left a vehicle bomb in the area, intending to maximize casualties after you have evacuated the building.
8. ***Designate search teams.*** The search teams should be utility employees, preferably employees who work in the area they are searching. They will know if something is out of place or suspicious. An outsider, *i.e.*, police officer or fire fighter, would not.
9. ***Designate areas to be searched.*** Part of your pre-incident planning is to identify locations where bombs would most likely be placed and to search these areas first.
10. ***Establish techniques to be utilized during the search.***
11. ***Establish a procedure to report and track progress of the search and a method to lead qualified bomb technicians to a suspicious package.***

12. ***Have contingency plan available if a bomb should go off.***
13. ***Establish a simple-to-follow procedure for the person receiving the bomb threat.***
14. ***Review your physical security plan in conjunction with the development of your bomb incident plan.***

Blueprint diagrams of the building should be kept in the primary and alternate command centers, both of which should be as far away as possible from locations where a bomb is most likely to be placed. There must be a means of communicating from the command center to the search teams *without using radios*. The use of a radio around an explosive device could cause a detonation.

The normal steps to a bomb threat are:

- ***A threat is communicated to the utility, usually by telephone.*** Bomb threats are also communicated using the mail, facsimile machines, or e-mail. If the bomb threat is made on the telephone, the person who receives it is responsible for getting as much information as possible from the caller. If the threat is received via the mail or fax machine, put it down; don't touch it until the police arrive. If it is received by e-mail, print a copy but do not attempt to save it as it may be accompanied by a virus. Some of the newer viruses do not require that you open an attachment, only that you save the e-mail to your computer. Leave it on the screen until the police arrive. The person receiving the threat notifies the team and law enforcement.
- ***The bomb threat response team evaluates the threat.*** It may coordinate the initial evaluation with police. The team will reach one of four decisions:
  1. Ignore the threat
  2. Do not evacuate, but conduct a preliminary search
  3. Evacuate immediately
  4. Evacuate and search

A bomb threat worksheet based on recommendations of the BATF is included as Appendix E. Copies of this worksheet should be at all locations where a telephone bomb threat could be received. The person receiving the call should try to keep the caller on the telephone as long as possible and to ask questions about the device, its location, and the motivation of the caller. A caller who can describe details of the device and the detonator will be taken more seriously than someone who cannot. Listen for background noises and accents or other unique speech characteristics. When the police arrive, they will interview the person who received the call.

Many utilities evacuate a site whenever a bomb threat is received. Depending upon the size of the building and the number of employees, this may be a prudent policy. A major utility evacuated 1000 people at its headquarters building five days in a row. A costly experience! The caller (who was identified) was a disgruntled employee who would leave the building and make the bomb threat call from a pay telephone. Initiating a search and evacuating, if a suspicious package is found, is an alternative to *carte blanche* evacuations. This is a decision only management can make.

In large buildings, other companies, government offices, *etc.*, may be located. This means that decisions to evacuate and other response decisions need to be coordinated with building security and management and with these other tenants. If an evacuation takes place in a multi-story building, evacuate by floor beginning with the suspect floor, then the floors above and below.

Each search team consists of two people who work in the area. Technique is key. For instance, assuming that the area has been evacuated, they walk into the middle of the room, close their eyes, and listen quietly for a minute or so. If a device has a clockwork mechanism, they may hear it. They divide the room into two equal parts (based on the size of room and the number of objects in it) and then divide into three levels of height. Each member searches his/her half of the room from about desk level down. This includes all packages, machines, and electric sockets or computer connection holes that may have been disturbed. They then search the area from desktop level to about head level. Again, they look for anything that does not belong in the area or that is otherwise suspicious. The third part of the search is from head level up, including the ceiling. Carefully examine drop ceilings or other openings where a device may have been placed. Look around light fixtures and other hanging objects. Check air conditioning and ventilation ducts and sound or speaker systems.

To summarize, the following steps should be followed in searching a room:

1. Stand in the center of the room, back to back, with your eyes closed and listen for at least 30 seconds.
2. Divide the room into two sections.
3. Start from the bottom and work up. Search at three levels, one at a time.
4. Start each search back to back, and work back towards each other.
5. Go around the wall, and then proceed toward the center of the room.

When the search is complete, close the door and place a piece of tape across the door and door jamb at eye level. Mark the tape, "search completed."

Under no circumstances should a searcher touch, jar, or move a suspicious device. If a device is located, report the location immediately to the search command center. Identify the danger area, and block it off with a clear zone of at least 300 feet, including floors above and below the object. Open all doors and windows in the area to minimize the damage from any blast and fragmentation.

BATF also provides recommendations for suspicious packages that arrive in the mail. Posters that illustrate the possible indicators of a suspicious package are available from BATF, the FBI, and the U.S. Postal Service. At least one of these posters should be in your mailroom. Here are the BATF recommendations:

1. If delivered by carrier, inspect for lumps, bulges, or protrusions, without applying pressure.
2. If delivered by carrier, balance check if lopsided or heavy-sided.
3. Handwritten address or labels from companies are improper. Check to see if the company exists and if they sent a package or letter.

4. Packages wrapped in string are automatically suspicious, as modern packaging materials have eliminated the need for twine or string.
5. Excess postage on small packages or letters that indicate the object was not weighed by the post office.
6. No postage or uncancelled postage.
7. Any foreign writing, addresses, or postage.
8. Handwritten notes, such as: "to be opened in the privacy of..." "CONFIDENTIAL," "your lucky day is here," "prize enclosed," *etc*.
9. Improper spelling of common names, places, or titles.
10. Generic or incorrect titles.
11. Leaks, stains, or protruding wires, string, tape, *etc*.
12. Hand delivered or "dropped off for a friend" packages or letters.
13. No return address or a nonsensical address.
14. Any letters or packages arriving before or after a phone call from an unknown person asking if the item was received.
15. If you have a suspicious package, *CALL 911, ISOLATE, EVACUATE*.

The anthrax mail attacks in the fall of 2001 made it clear that bombs and other explosive devices are not the only threat to mailrooms. Utilities should have a general mailroom security plan and procedures that apply to the handling of all mail. However, there is a major difference in the response to a suspicious package or letter that may be a bomb and one that may contain a biohazard.

If the suspect device is a bomb, everyone should evacuate the area; if the suspect letter or package may contain a biohazard, then it should be isolated within the mailroom, but everyone present should stay there. If the people who handled the mail leave the area, they may contaminate a larger area. Place the suspect letter or package in a biohazard bag (available from a safety supply company), and wash your hands and other exposed skin. If there is a concern with regard to a biohazard threat, employees handling the mail should wear disposable gloves and protective masks. These items can be purchased from a safety supply store. Masks are either N-95 or N-99. (Although the N-95 meets the recommendations of many agencies, spend a little more and purchase the N-99 masks.)

For more information on mailroom security, contact your local postmaster and ask for a copy of *Mail Center Security Guide, Publication 166*. The guide is also available on-line at *www.usps.com/cpim/ftp/pubs/pub166/welcome.htm*. The Centers for Disease Control (CDC) has a video available, *Protecting Your Health: For People Who Process, Sort, and Deliver the Mail*. The video is free and is available at *www.osha-safety-training.net/ANT/anthrax.html*. If you cannot find the video at this site, do a key word search at the CDC home page, *www.cdc.gov*.

Executive Protection

Look at your utility's web site. Is there a photo of the CEO or general manager? If so, suggest that it be removed. Why provide an angry customer with a photo that can be downloaded and used for target practice, before going after the individual?

Major threats to executives come from angry customers, disgruntled ex- or current employees, extremists, and two-bit criminals. Threats include kidnapping, murder, bombing, arson, and extortion. In general, threats from each of these can be summarized as follows:

- Terrorists or other extremists:
  - extortion
  - bombing
  - kidnapping
  - arson
  - assassination
- Organized crime:
  - extortion
  - kidnapping
  - murder

- Two-bit criminals:
  - robbery
  - extortion
  - kidnapping
  - contract murder
- Disgruntled (ex) employee:
  - hostage taking
  - extortion
  - kidnapping
  - murder
- Mentally disturbed individuals:
  - bombing
  - attack: shooting or attack on vehicle
  - kidnapping
  - murder

In executive and dignitary protection, the potential target is referred to as the principal. The primary job of the security detail is to prevent the assassination or intentional injury of the principal. They are also responsible for preventing unintentional injury and embarrassing events.

Executive protection of high-threat targets requires trained personnel. This is not a job for amateurs! For most utilities, the protection of the CEO or general manager is less challenging. Again, the amount of security needed depends upon the level of the threat. If the utility is going through major downsizing, the threat to executives may be high. If it has received considerable negative publicity, the threat is elevated. As with other security measures, the time to plan for increased security is before it is needed.

Some of the basic things your utility can do to protect its key executive and managers include the following:

- ***Access to executive/management offices should be controlled.*** This includes restroom facilities.

- ***A security survey should be conducted of key personnel homes.*** Adequate physical security and access-control measures should be in place. Install alarm systems at these homes. These should be paid for by the utility.
- ***Family members must be involved in the security process.*** They should report any suspicious people or activity near the home and be alert to suspicious mail.
- ***During increased threat levels, the executive should carry a panic alarm at all times.*** If the alarm is activated, the security company will notify the police. Executive spouses may also need this protection.
- ***Consider maintaining an executive crisis response file on each key executive.*** These will be needed if a kidnapping or hostage incident occurs. An example of this file is included as Appendix G.
- ***Develop a code system.*** This will be used if the executive is calling the office or family. If he/she asks a certain question, it can be an indication that something is wrong.

When traveling, executives should have specific times when they call the office or family. Their itinerary should remain as confidential as possible, especially hotel and restaurant information. They need to take responsibility for their security and safety. If something looks wrong, it probably is. Don't enter suspicious vehicles or leave the hotel if there is questionable activity outside. Keep a low profile.

If an event does occur, the utility's primary responsibility is the protection of the family. Law enforcement will be responsible for responding to the incident itself. The family should be moved to a secure location, unknown to the media, and provided the support they will need until the situation is resolved.

The best security is prevention. Principals should receive periodic briefings on current threats and be reminded of the security procedures they should follow. Some take these threats seriously and work with security. Others are more challenging.

Extortion

The target of an extortion attempt may be an executive or the utility itself. An executive may be threatened with physical harm to him/herself or family if certain demands are not met. The utility may be threatened with a bombing or an attack on its network. Extortion attempts, like bomb threats, may be received on the telephone, in the mail, via facsimile, or by e-mail. If the threat is received on the telephone, get as much information as possible. If it is received by mail, fax, or e-mail, follow the same procedures as for a bomb threat.

Call the police immediately. Depending upon the nature of the threat, it may be a violation of state or federal statutes. Preserve whatever evidence is available, and call the police.

Incidents of cyber-extortion are increasing. If your utility is also an ISP, you may be targeted. Hackers break into systems and steal credit card information on subscribers. They then threaten to release the information on the web unless the ISP meets their demands. This has been an embarrassment for more than one ISP.

Another cyber-extortion plot has the hacker break into your system and threaten to post the methods used on hacker bulletin boards unless the company pays a "consulting fee." Two Kazakhs, Oleg Zezov and Igor Yarimaka, attempted to extort $200,000 in "consulting fees" from Bloomberg LP. They e-mailed the CEO of the company a copy of his own security badge to prove they had compromised his network security. Michael R. Bloomberg, CEO, agreed to meet with the pair and brought two police officers posing as Bloomberg employees. The extortionists have been charged on three felony counts including extortion and unauthorized computer intrusion.

In some parts of the world, extortion is a major activity for terrorist groups. When the police raided a "safe house" belonging to the Tupac Revolutionary Movement in Peru, they found computer records identifying companies that were currently paying extortion money to the group and companies they planned to extort in the future. The Irish Republican Army and the Loyalist movements in Northern Ireland were both involved in extortion. This is also a major activity of organized crime groups around the globe.

Here is the bottom line on dealing with extortion:

- If the threat is made on the telephone, get as much information as possible. Write everything down.
- If the threat is received via mail, fax, or e-mail, follow the same procedures as for bomb threats.
- Call the police immediately.
- If certain individuals are threatened, make sure they get to a safe area.
- Work with the police to resolve the case and apprehend the extortionist.

This is not a major problem in the utility industry—at least, not yet. You should be prepared to respond, especially if you provide ISP services or have customer credit and credit card information stored in your network system.

Theft-of-Service and Customer Fraud

In the past, many utilities have chosen to ignore the theft-of-services problem, but, as competition becomes a reality and customers and regulators gain an understanding of the potential extent of the problem, revenue protection has become increasingly important. It can also be highly embarrassing if a case that the utility should have prevented is detected.

At one large utility, the police arrested a *fixer* who was rolling back dials on electric meters. He had more than 200 monthly accounts that he would visit; each paid $25 per month to have their dials adjusted. What does this tell you about the utility's seal program? It was nonexistent! They not only were improperly using external seals but were not using internal seals either. The arrest made the front page of the local newspaper. The utility now has an excellent revenue recovery program. The fixer skipped bail and is probably rigging meters in another city.

You don't want your utility on CNN, especially if they are reporting a story as they did in February 2001. A customer had been stealing from the utility since he got out of the military at the end of World War II. The case was discovered when the customer, now 91-years-old, called to complain about an outage. The utility back-billed the customer for just three years—$82,000—which means that at today's rates, this customer stole approximately $1.5 *million* from the utility, at a single residence.[1]

Theft-of-service results in loss of lives as well. A customer who kept reconnecting his electric service when he was disconnected for nonpayment finally had the fuse removed from the transformer in front of his house. He acquired a fuse and climbed the pole to refuse the transformer. He then closed the disconnect switch with his hand. He is what is known as a *no-repeat offender*.

Developing a revenue protection program results in the recovery of revenues from losses other than theft and fraud. When employees become more aware of metering, meter reading, and billing, they also discover metering and billing problems. At one utility, recoveries total approximately $300,000 per year in meter tampering and fraud cases. They recovered more than $900,000 in a single case involving improper metering. The industrial metering had not been installed correctly, and they lost $300,000 per month for three months. (The revenue was recovered.)

When the International Utility Revenue Protection Association (IURPA) was formed in the late 1980s, the emphasis on revenue protection was primarily on theft-of-service. As members from around the world came together, they found that customer fraud was another common problem affecting the bottom line of most utilities. Theft-of-service may cost a utility from .005% to .025% of gross revenues, but fraud losses could be three or four times greater. For more information on IURPA, visit their web site, *www.iurpa.org*.

Theft-of-service also includes customers who fail to sign up for service and connect themselves, active customers who steal a portion of the service and pay for the rest, and customers who have been disconnected for nonpayment and reconnect the service. These thieves either tamper with or bypass the meter.

Theft-of-service is a form of customer fraud. Other fraud losses include the following:

- Fraudulent applications for service
- Customers who leave owing the utility money
- Bad checks
- Revolving door accounts (where people at the same location change the name on the service after being disconnected for nonpayment)
- "Out of business" / "back in business" (businesses that intend to fail and then open under another name)

Outside employees are key to detecting theft-of-service and contribute to the detection of revolving door fraud. These employees need to know how to recognize a service where tampering may have occurred (or is obviously occurring), and there must be a system in place for reporting the situation.

Tampering will also be detected by the monthly exceptions report. The computer-generated report lists the accounts in which usage is higher or lower than expected. The formula used for many of these reports is simple—compare this month to the same month last year. Others are more complicated—compare the average usage for the past 14 months, but take out the highest and lowest months. In tropical climates, the report may compare month-to-month usage.

When a theft is detected, it must be investigated and evidence collected according to the rules of evidence. All of the evidence found at the service must be documented, tagged, and bagged. It must then be locked in a secure area at the utility. Photograph the evidence as it is found at the scene, enter a description of it in your field notes and final report, put an evidence tag on larger items such as meters, and place smaller items (such as seals) in evidence bags.

The primary objective of most revenue protection programs is recovery of lost billing. Once the bill is estimated, the utility should contact the customer, present the evidence and the bill, and collect the lost revenues.

If the customer is not cooperative, the case should go to court. If a customer is caught a second time, or if you identify a fixer, the repeat offender and/or fixer should be charged with violation of the appropriate laws in your state or province.

The best method for reducing losses resulting from theft-of-service is to make it more difficult for thieves to commit the crime. The utility must have an aggressive sealing program, and seals on meters should be inspected on a regular basis—at least annually. This is especially important if the system is using automatic meter reading (AMR) systems. Most AMR systems detect and report meter tampering. But, as some utilities are learning, when these systems are installed and customers learn that the meter reader won't be coming around anymore, they save on their utility bills with bypass systems that do not trigger the AMR alarms.

Detecting and controlling fraud losses requires a good customer records system. It is amazing to learn how quickly customers move from location to location. If customer A is disconnected at 10:00 this morning for nonpayment, customer B will move into the location and request new service that afternoon. This is, in most cases, a revolving door account. The service should not be connected until the identity of customer B and the "new" address of customer A is verified. If the service employee arrives at the location to find the same cars, patio furniture, etc., don't connect the service.

In many cases, an inexpensive credit check will verify the information provided by the customer. If there is a problem, the customer needs to come to the utility office and verify the provided information before service is connected. When you identify someone who skipped on you in the past or is playing the revolving door game, assume that they have done this to you at other locations. Utilities that investigate these cases report that the $300 to $400 average loss at the location where the fraud is detected becomes a back bill closer to $2,000 when they identify the locations where the customer has defrauded them in the past.

As with other types of security threats, the utility should have a revenue protection policy. An example of a policy is included in Appendix H.

Summary

As with any category of business, utilities experience special threats such as bomb threats, threats to key individuals, and the potential for extortion. And, like any other business, there are customers who decide to steal rather than pay for your product. Your security program should include policies and procedures to address all these threats.

Bomb threats can have a detrimental effect on business operations and employee productivity unless you are prepared to deal with them before the threat is received. Responding to a bomb threat requires prior planning and coordination with police and fire departments. It may require that the utility designate and train search teams. These activities need to happen before the threat is received—not after.

The same is true when responding to physical threats to key executives and extortion attempts. Liaisons with law enforcement and procedures for protecting threatened employees and their families should be in place. The first telephone call the utility should make after receiving a threat or an extortion attempt is to the police.

Theft-of-service and customer fraud is taking money from the utility's bottom line. Even worse, unless these losses are addressed, they eventually are built into the rate base, and all customers pay for what a few are stealing. Every utility should have an aggressive program to address this program and to minimize these losses.

Action Checklist

1. Develop a bomb threat response plan.
    a. Designate a chain of command and a bomb threat response team. If in the United States, contact the local office of the Bureau of Alcohol, Tobacco, and Firearms. They have pamphlets and other materials that will assist in organizing and training the team. If in another country, contact your local law enforcement agency.
    b. Establish a command center and alternative command center. There should be a complete set of building floor plans at both locations.
    c. Develop an evacuation plan. Try not to evacuate to parking lot areas.
    d. Train personnel to respond to bomb threats. If received on the telephone, they should complete the information in Appendix F.
    e. Train your bomb search teams.
2. Prepare for executive protection.
    a. Identify executives who are most likely to be targeted.
    b. Ensure that access control procedures are in place in the areas where these people work.
    c. Develop and secure executive crisis response files for each of these executives. See Appendix G.
    d. Improve the security at their residences. This includes the use of alarm systems.
    e. Encourage executives to follow appropriate security procedures when traveling.
3. Be prepared to respond to extortion attempts.
    a. Train employees to get as much information as possible if they receive an extortion attempt on the telephone. Train them to handle the evidence correctly if it is received by mail, fax, or e-mail.
    b. Call the police immediately.
    c. Assess the possibility of the utility being the target for cyber extortionists. If at risk, make sure the controls are in place to minimize the threat.

4. Review your revenue protection procedures.
   a. Seal all meters, and check seals on a regular basis, at least once a year.
   b. Train field personnel to detect and report tampered service and bypasses.
   c. Establish and review the monthly exceptions report.
   d. Use a credit reporting service to verify information on new customers, skips, and problem customers.
   e. When theft or fraud is detected, meet with the customer and attempt to collect the amount past due. If the customer is not cooperative, consider using civil courts to sue them.
   f. If a customer is caught a second time, or if you identify a fixer, contact the police or the prosecutor's office. These people should face criminal charges.

Notes

1. Seger, Karl A., "Energy Theft: An International Perspective," *Metering International*, Issue #1, 2002

10

Threat Response Plan

The Infrastructure Threat

PDD 63 and the NERC *Alert Levels and Physical Response Guidelines* were discussed in chapter 1. The guidelines are found in Appendix A.

The objective of this chapter is to assist utilities in developing a threat response plan based on the NERC guidelines. The guidelines are based on the same approach used by DOD, FAA, and many other organizations. The security measures implemented are commensurate with the level of threat.

Several points should be made:

- ***Terrorists target infrastructure.*** Ahmed Ressam, an al-Qaida operative who was convicted of conspiracy to bomb the Los Angeles airport around Y2K, stated in court that he had been trained at a camp in Afghanistan to target airports, hotels, military installations, tourist sites, embassies, warships, and infrastructure targets—including utility systems.

- ***When an infrastructure target is attacked, there is a cascading effect.*** If the electricity in an area is disrupted for a considerable amount of time, there is an impact on water and wastewater treatment, medical facilities, and a number of other services vital to the community.
- ***The potential impact of these cascading effects are generally underestimated.*** Due to a lack of understanding with regard to the interdependencies of infrastructure targets, many of the components involved do not comprehend the cascading effects of an attack on infrastructure and are not prepared to respond if a major event does take place.

At least two major desktop exercises assess the cascading effects of attacks on infrastructure targets. The Black Ice exercise, coordinated by the Utah Department of Public Safety's Division of Comprehensive Emergency Management, the U.S. Department of Energy's Office of Critical Infrastructure Protection, and the Utah Olympic Public Safety Command, created a "disaster-resistant Olympics." Another major exercise, Blue Cascades, identified potential cascading effects of an infrastructure attack in the Pacific Northwest. U.S. and Canadian organizations participated. A key coordinator for both exercises was Dr. Paula Scalingi, formerly director of the DOE Office of Critical Infrastructure Protection and currently president of the Scalingi Group.[1]

Approximately 225 people, representing more than 65 organizations, participated in Black Ice. These included players from the public and private sectors. Key issues addressed during the exercise were interdependencies, coordination, communications, and resource allocation.

One of the needs identified during Black Ice was for a better understanding of what interdependencies really are and their importance to the operation of infrastructures. These include an understanding of interdependencies during normal operations as well as during disruptions. Other needs identified included better interaction between infrastructure service providers and others in the community and the need to identify and understand SCADA system vulnerabilities and interdependencies.

Coordination needs were identified, including ensuring close and continuing coordination among all of the infrastructure service providers and establishing procedures for maintaining this coordination during disruptions. Other coordination needs included establishing procedures for prioritizing and coordinating service restoration and avoiding "who's in charge?" problems. A final need was coordination with the media and the public during major disruptions.

The need for effective regional communications among the providers and other stakeholders during normal operations was discussed. The need to identify and test primary and back-up systems for both normal and disruptive operations and the need to develop plans to maintain effective communications during system overloads were addressed.

One of the key resource allocation issues identified was the need to get the right people to the right locations during infrastructure disruptions. Other resource allocation needs included coordinating for Olympic operations and emergency responses and the need to identify and establish priority resource allocations.

An outcome of Black Ice was development of a set of recommendations on how to create a "disaster-resistant" state or region:

- Establish a stakeholder group; if possible, expand existing coordination mechanisms [this should include all infrastructures, federal, state, and local (*e.g.*, police, fire) government, communities]
- Exchange general information (hold briefings) on infrastructure operations, interdependencies, response/recovery plans
- Develop a regional infrastructure assurance implementation plan that includes:
    - an inventory of information capabilities
    - a threat analysis
    - infrastructure vulnerability assessments
    - community vulnerability assessments
    - enhancement of response/recovery plans
    - training and exercises
    - identification of necessary databases, analytic tools, equipment
    - resources needed, and source of resources
- Hold infrastructure interdependency exercises to help develop the implementation plan, to identify key shortfalls, or to test protection, mitigations, and response enhancements

- Develop a task and milestone schedule
- Identify "champions" for each task
- Get expert help as necessary

Many of these same general needs were identified during Blue Cascades. Cascading effects of a disruption to a region's electric power would cause outages, which would quickly spread to other states and result in the disruptions of the region's telecommunications, natural gas distribution, and municipal water systems. Other services impacted would include ports in the region, transportation, emergency services, hospital mass care, and law enforcement.

A utility's threat response plan must be coordinated with other infrastructure response plans at all levels, *i.e.*, federal, state, regional, and local. But before you begin to undertake this challenging effort, develop a threat response plan that addresses the utility's specific needs, and then integrate your plan into other infrastructure threat and incident response plans. The interdependencies of critical infrastructure components and the cascading effects of a disruption to any component must be considered and included in potential scenarios and responses.

The DOD Model

Few organizations have been confronted by more different forms of terrorism, extremism, and other security challenges than the U.S. DOD. Your utility is probably not protecting itself against the same degree of threats, nor are you preparing to go to war, but many lessons can be learned from the DOD approach.

Terminology is different, but the approach is basically the same as our risk formula: *Risk = Threat + Criticality/Vulnerability*. During the threat assessment stage, risk assessment graphs are developed to identify incidents with the greatest likelihood of occurrence and greatest consequence should they occur. A similar approach is displayed in Figure 10–1.

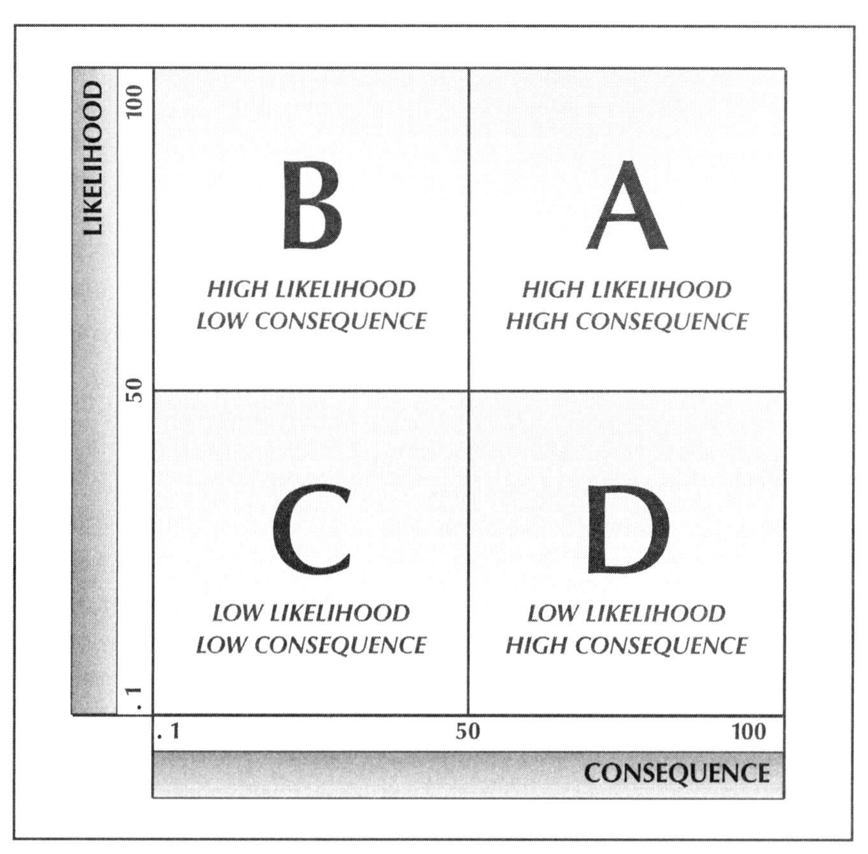

Fig. 10–1 Risk Assessment Chart

The risk assessment chart has four quadrants:

Quadrant A: high likelihood and high consequence
Quadrant B: high likelihood and low consequence
Quadrant C: low likelihood and low consequence
Quadrant D: low likelihood and high consequence

When you conduct your risk assessment, determine the likelihood that a particular event could occur on a scale of 0 (not likely to occur) to

100 (most likely to occur), and determine the potential consequence for the event on a scale of 0 to 100. Place the event on the risk assessment chart. For example, if there have been a number of major vandalism events targeting substations in your area, then the likelihood of an event occurring is high—let's say, 80 points. One of your transmission substations is located in a rural area but near a highway. You determine that the consequence of this substation being vandalized would be high—also 80 points. The event would be entered into Quadrant A at the 80-point intersection. This is a Quadrant A event—high likelihood and high consequence. The threat assessment committee should complete this exercise for each likely event, and then enter their findings into the risk assessment chart. These are "guesstimates," but this is an effective approach for identifying threats that must be addressed.

It is impossible to prepare a response to every potential event. Place the emphasis on events in Quadrant A. Next, focus on the events in Quadrant C. The events in Quadrant B have a high probability of occurring but a low consequence. Events in Quadrant D should be identified, but few utilities have the additional resources needed to address them. If you are prepared to respond to threats in Quadrants A and C, then you will be able to deal with any event in Quadrant D.

DOD Regulation 5200.8-R defines the assessment used to determine the threat condition (Threatcon) and level based upon the known threat. The DOD threat levels range from Threatcon Alpha (minimum threat) to Threatcon Delta (terrorism is occurring). This is the same approach used in the NERC plan, although the determining criteria are different.

Action Set Matrix

The NERC threat alert levels and physical response guidelines are designed to be implemented at national, regional, and local levels. Currently, national alert levels are determined at the NIPC and are based on specific threats to the utility industry. *Threatcon ES-Physical-Green* applies when no known threat exists. We will not return to this level for years. *Threatcon ES-Physical-Red* applies only when an incident occurs or creditable intelligence indicates that a terrorist or criminal act against the industry is imminent or has occurred. You must be prepared to implement *Threatcon ES-Physical-Red*.

The NERC guidelines in Appendix A are used as follows:

1. The security or threat management committee is responsible for the development of the plan. The committee should include operations and security professionals. A separate plan may be needed for each critical facility. A general plan may be developed for all noncritical facilities.
2. You should be receiving threat condition information from the NIPC. They publish an electronic report, *NIPC Daily Open Source Report*, which provides important threat condition information. For example, when the White House determined that the overall terrorist threat to U.S. assets at home and overseas was elevated, *Threatcon ES-Physical-Orange* (the second highest level) was implemented. For subscription information, contact the NIPC Watch and Warning Unit (202.323.3204 or **nipc.watch@fbi.gov**).
3. Begin with *Threatcon ES-Physical-Green (low)*. There are three recommended security procedures listed. The security or threat management committee will list other basic procedures that should be in place at all times. Be specific. Which doors should always be locked? Which areas should always be monitored? Remember, you may have to have a different plan for different facilities.
4. *Threatcon ES-Physical-Blue (guarded)* is the level utilities will be most of the time. At this level, all of the measures listed under *Threatcon ES-Physical-Green* are implemented, plus additional measures that improve the utility's basic security. The example lists measures 4 and 5; add to these.
5. NERC recommended procedures 6 through 10 are listed under *Threatcon ES-Physical-Yellow (elevated)*. You should already have measures 1 through 5 and the others you added to the previous Threatcon levels. The committee will add additional procedures that are specific to your utility.

6. At *Threatcon ES-Physical-Orange (high)*, NERC recommends that procedures 11 through 24 are activated. Business is not going to be normal at this level. Customer service facilities will be closed. Nonessential employees may be sent home, and essential employees are either manning specific locations or on call. Your critical facilities are on lock-down, and security guards may be in place.
7. Hopefully, you will never go to *Threatcon ES-Physical-Red,* but you must be prepared. Few people believed that September 11, 2001, could possibly happen. But it did.
8. The "Additional Threatcon Response Considerations" lists additional areas to consider when developing your Threatcon plans. Again, the committee should identify additional areas that are specific to your operations.
9. Copies of the plan are maintained at each facility. Since the information is sensitive, it should be labeled "Company Confidential" and kept in a secure area.
10. Train people to implement the plan. This includes the key players and their backups. Everyone who may be responsible for implementing a Threatcon security procedure or for ensuring that the procedure is implemented participates in the training.

Making the Plan Work

You have been at *Threatcon ES-Physical-Yellow,* and suddenly *Threatcon ES-Physical-Orange* is activated. Who implements the additional security measures listed in your plan, and how do we ensure that they are implemented? Organizations that have experienced major threats have found that having a plan doesn't mean it will be implemented when needed or that it is adequate to address the security needs at each level. You must identify persons responsible for implementing the procedures and designate channels to ensure that they are implemented. The plan must also be tested periodically.

The action set matrix helps ensure that measures are implemented or that they cease as threat levels change. There must be an action set matrix for each of the security procedures listed in your plan. This may seem like a daunting task, but the plan is not complete until the action set matrixes are developed.

Each matrix must include:

- Threatcon level
- Number of the Threatcon response
- Specific action(s) required
- Individual responsible for the action
- Person to whom completion of the action is reported

When identifying the individual responsible for the action and the person to whom they report, do not use specific names. Use titles or positions. "Nick Jones, the office manager," may not be on site to lock the back doors when the threat is elevated. In that case, whoever is the acting office manager will complete the task.

The final page of Appendix A is an action set matrix for Threatcon measure 7 in the NERC guidelines. The guidelines do not include information on this procedure. It was adopted from the *Anti-terrorism Force Protection Installation Planning Template* published by DOD. DOD has learned from experience that, if a Threatcon plan is going to work when it is needed, it must include an action set matrix for each procedure. In the words of Mark Twain, "A smart man learns from his own mistakes, but a wise man learns from the mistakes of others." Learn from DOD's experience, and develop a complete threat condition plan that includes action set matrixes.

Management of a major West Coast utility decided to update and test its emergency response plan using an earthquake scenario for the test. A month later, they experienced a devastating earthquake, and, as a result of the exercise, they were better prepared to respond. Damage was quickly assessed, most employees knew how to respond and where to report, and services were restored in record time. The utility also did an excellent job of developing a Lessons Learned Report after the event and implemented these lessons into its response plans.

A plan is never complete until it is tested or exercised. There are three basic approaches to accomplishing this objective: what-if analysis, desktop exercises, and field training exercises.

An extremist group just bombed the facilities of a utility on the other side of the country. This is an opportunity for your threat assessment committee to meet and ask, "What if they tried the same thing here?" This exercise will help to identify and update threats and response measures. It also helps to develop the group dynamics and the decision-making ability of the committee, so when your utility is directly threatened, you are better prepared to analyze the situation and make appropriate recommendations to management. If you don't know of any incidents that have occurred, the committee could meet periodically and discuss potential scenarios. This could be a quarterly lunch meeting.

The scenarios discussed could be developed into a desktop exercise. All key players in the utility should participate in the exercise, which should last for at least half a day. Emergency response or law enforcement personnel should be present to facilitate and evaluate the exercise. The exercise will identify holes in your plans and help everyone to better understand their roles if the plan has to be implemented.

Some communities have community-wide field training exercises. Major cities in the United States are conducting exercises that involve weapons of mass destruction. In regions with mines, mine evacuation drills are often conducted. As the Homeland Defense program continues to evolve, more of these exercises will take place at regional and local levels. Utilities, as a key component of the infrastructure, should participate in these exercises. If these exercises are going to be beneficial, then actual players should participate. These are the people who will make the decisions if something actually happens. If participation is delegated to someone else, it shows a lack of commitment and interest on the part of the person who should have been there. It also means that the key players may be less prepared if a major incident does occur.

"What-if" exercises and desktop exercises can be conducted by the utility without other infrastructure components. Your utility could take the lead in the community, however, by inviting other organizations to participate. It would help to gain a better understanding of the interdependencies and the potential cascading effect if a major event ever takes place. It will help you to identify coordination, communication, and resource needs as they were identified during Black Ice and Blue Cascades. It will also give you the opportunity to share Threatcon response plans so that the community as a whole is better prepared to respond to an event.

Summary

The threat condition program and recommended responses will continue to evolve. The future of the NIPC and its relationship to the Department of Homeland Defense has not been decided. The role that NERC and other utility organizations and associations will fill is unclear.

What is clear is that the threat remains. Utilities, more than many other industries, had disaster plans in place before September 11, 2001. These plans need to be expanded to include the new security threats, including infrastructure targeting. They should, as NERC recommends, have five levels of operation. If they include a set of action set matrixes and if they have been tested, then your utility is as ready as it can be for a worst-case scenario.

Action Checklist

1. Consider taking the lead in your community in developing a program to identify infrastructure interdependencies and potential cascading effects. Use the outcomes of Black Ice as a model.
   a. Establish a stakeholder group.
   b. Hold briefings, and exchange general information.
   c. Develop a regional infrastructure assurance implementation plan.
   d. Hold infrastructure interdependencies exercises.
   e. Develop a task and milestone schedule.
   f. Identify "champions" for each task.
   g. Get expert help as necessary.

2. Develop a risk assessment chart.
   a. The threat committee identifies potential events and estimates their probability of occurrence (scale 0-100) and potential consequence (scale 0-100).
   b. Chart these events on the risk assessment chart.
   c. Develop prevention and response plans for events in Quadrants A and C.

3. Develop a threat response plan based on the NERC guidelines.
   a. Subscribe to the *NIPC Daily Open Source Report*.
   b. Develop measures, in addition to the NERC recommendations, for each threat condition level.
   c. Review the "Additional Threatcon Response Considerations."
   d. Develop an action set matrix for each response measure.
   e. Train people to implement and use the plan.

4. Test your plans and procedures.
   a. Conduct regular "what-if" discussions. If a major event occurs, have a special "what-if" meeting.
   b. Conduct desktop exercises at the utility. Consider inviting others to participate.
   c. Participate in regional and community desktop and field training exercises.

Notes

1. Verton, Dan, "Exercises Expose Vulnerabilities," *Computer World*, July 8, 2002, *http://www.computerworld.com*

11

Crisis Management

Learning from Experience

Utilities are ahead of most organizations on the learning curve when it comes to responding to a disaster. Many of their experiences in responding to natural disasters can be applied in preparing to respond to a major security incident.

In chapter 10, we discussed a utility that reviewed and exercised its emergency response to earthquakes and then experienced that event in real life a month later. This was PG&E. Their emergency operations center was prepared. Employees knew where to report, and if they were not able to reach their assigned location, they knew to go to an assembly area they could reach. Methods were in place to implement emergency operations, assess the damage, and implement damage control procedures. Natural disasters are going to happen; most utilities are prepared for the types of disasters that impact their area of operations.

When an earthquake struck Southern California on January 17, 1994, causing $30 billion dollars in damage, the Southern California Gas Company learned how important their crisis response plans were. The company had an experienced crisis response team in place, and they had conducted a vulnerability analysis prior to the earthquake. They had conducted training, developed emergency coordination plans, and made a point of learning from previous events.

When the event occurred, the emergency operations center was activated in 12-hour shifts, and a damage assessment was conducted immediately. During the next five days, 2940 employees concentrated their efforts in the most affected areas. They were supported with emergency supplies stocked at the company's satellite offices and provided special hours, housing, and counseling as needed. The company communicated with its customers through the media, recorded telephone messages, and with door-to-door flyers.

The company listed seven important findings in the incident Lessons Learned Report:[1]

1. The value of a written response plan.
2. The importance of working with government agencies, especially Federal Emergency Management Agency (FEMA).
3. The need to assign equipment on a priority basis.
4. Sharing of information.
5. The *emergency operations center* (EOC) must have a master list of home and other contact numbers.
6. The need for additional communications equipment.
7. The value of working with the media.

When developing an emergency operations capability, there are three important considerations: the emergency operations plan, the emergency operations center, and the emergency operations team.

Con Edison had to respond to the devastating attack on the WTC on September 11, 2001. The company lost two substations when the towers fell. More than 1900 Con Edison employees worked around the clock, and, within days, they had laid more than 33 miles of high voltage cable, restored power to part of the area, and provided generators to help other critical customers. The lessons learned in this response are still being evaluated. The utility and its employees deserve the highest praise.[2]

Emergency Operations Plan

The initial consideration of the emergency operations plan addresses authority and jurisdiction. (Who within the utility will be in charge, and which law enforcement agency will have jurisdiction?) If the event is criminal, *e.g.*, hostage taking during a robbery, then local law enforcement will have jurisdiction. If this is declared a terrorist event, then the FBI will assume jurisdiction. The FBI has jurisdiction over most terrorist events in the United States. Other federal agencies have jurisdiction over terrorist events in other countries. The emergency operations plan should state who within the utility is responsible for managing events, but it should also recognize which law enforcement agency will have jurisdiction over these events. This will determine the notification procedures for various categories of events.

Notification procedures should include the notification schedule. (Who is notified first, *etc.*?) As was stated in the Southern California Gas "lessons learned," notification centers need to have all available contact telephone numbers for personnel and agencies to be contacted in an emergency. These numbers should be updated at least once a year.

The plan identifies mutual-assist contacts, including other utilities, government agencies, and vendors. It should state the type of assistance they can provide and the procedures used to request this assistance. Within the utility industry, many national, regional, and state associations maintain mutual assistance databases. It may be necessary to have a pre-existing memorandum of understanding or other agreements to activate mutual assistance when needed.

The emergency operations plan should outline the role of personnel responding to the event and the training they should receive. This part of the plan should include the following:

- ***Personnel response plan*** Who will be responsible for what, and who will perform each job and function?
- ***Contacts at government agencies for liaison during an event*** These contacts may also provide training and consulting assistance.

- ***Dedication to realism*** Establish the requirement that the "real players" will participate in all training, both desktop and field training exercises.

The plan should outline operational considerations. These are used to prioritize the response to an incident. It must also consider the impact of employee bargaining on emergency operations and your utility's relations with the media.

Specific responses at each of the utility's locations are part of the plan. These are the response procedures listed in the NERC plan. You will never know if they are actually implemented unless you include the action set matrixes as part of your plan.

The plan should address the need for additional resources in response to an incident. When confronted with inevitability of Hurricane Hugo, utilities in South Carolina began ordering all of the materials they would need for recovery after the storm. As the storm approached from the east, these materials were already in transit from the west. Recovery time was significantly reduced following the disaster, the results of which required generation to be re-established at major generation facilities from a cold start.

Your personnel will be on emergency operations for days or weeks. Where do they get the necessities of life, including food and fuel? As many utilities have learned, gas pumps don't work when the electricity is off (a cascading effect). Crews from mutual response organizations need to be housed and fed. They need to have their clothing washed, and they need a means to call their families. These needs must be considered prior to the event.

Your emergency response plan must outline the procedures to be followed at each of the five stages of a crisis:

1. ***Immediate response*** Your mission is to contain, confirm, and stabilize the situation.
2. ***Secondary response*** Conduct damage assessment and gain control over the situation.
3. ***Damage control*** Begin recovery of operations.

4. ***Resolution*** Restore operations to pre-incident conditions.
5. ***Lessons learned*** Conduct this analysis for internal purposes, but then share it with others in industry publications, at meetings, and in correspondence with other utilities.

To summarize, your basic response plan should include the following:

1. Basic crisis management team organization—be specific
2. Allocation of personnel, equipment, and other resources
3. Contact information for all personnel and outside resources
4. Immediate responses and decisions to be made
5. Plans for delegation and management of the incident
6. Notification and communication plans
7. Media and public communications plan

All employees must be a part of number 7. A field employee trying to restore service in an area will be approached by the media. A customer service employee answering telephone calls will get a call from a reporter. In both cases, the employee should not provide any information directly to the media. The media should be referred to the public information officer. Neither employee has the information needed to provide an assessment of the big picture and may say something the utility will regret later. All media points of contact are referred to one central location. This is what the U.S. Army's public affairs program refers to as, "speaking in one voice."

The callout list maintained in the plan should include the following contacts:

- The utility's callout list
- Law enforcement agencies (local, state, and federal)
- Fire departments and other emergency response organizations
- Local and state government officials
- Other federal agencies, including FEMA
- Mutual response organizations

- Organizations with specialized equipment
- Hospitals and other medical facilities
- Disaster response agencies, including the Red Cross and the Salvation Army
- Support groups such as amateur radio contacts
- Local media

The EOC

The EOC must be at a secure location. There should be an alternative EOC, also located at a secure location. In a large utility, the EOC may house several hundred people during a major event, whereas in a smaller utility, it may be 10 or 20 employees. In either case, there must be provisions to feed and provide other basic needs during the duration. Expect the EOC to operate on a 12-hour schedule, which means the utility must be prepared to provide for the needs of two teams during the crisis. Although two teams may operate in the EOC, there is only one boss. Have a separate place to house the CEO or manager, and try to maintain this person on the 12-hour shift schedule so that he/she can have some rest. But if something untoward occurs during the manager's off-hours, this person will be notified and will take charge of the situation.

In some crises, it may not be necessary to activate the entire EOC. If a hurricane, earthquake, or infrastructure attack occurs, the entire EOC will be activated. If the incident is isolated to one or two facilities, there may be a partial activation of the team. Your crisis response plan should state which events will require an initial partial activation of the EOC and which events require an immediate total activation of the EOC.

Use the FEMA approach; plan for the worst-case scenario.[3] When you do so, you will find that you are not prepared to respond to it, and you hope it never happens. You will, however, be better prepared to respond to other events as they occur. The worst-case scenario means you will have at least two shifts operating the EOC for an extended period of time. Remember Hurricane Hugo, when it took a month to get services back to somewhere near normal operations. The same could happen to your utility as a result of a natural disaster or an infrastructure attack.

The basic functions of the EOC include:

- Coordination of the response effort
- Proactive policy making (making decisions before the event occurs)
- Responsibility for ongoing operations
- Coordination of information collection
- Releasing public information through the media ("speaking in one voice")
- Hosting visitors to the EOC

The key roles of personnel in the EOC include:

- Emergency operations manager (the CEO or general manager)
- Operations specialist
- Logistics specialist
- Information specialist who monitors the situation
- Communications specialist
- Public information officer
- External liaison specialist (government agencies)
- External liaison specialist (mutual response organizations)
- Legal advisor

The EOC team must be prepared to work with the media during a crisis. If the crisis is an infrastructure attack or other crime, the lead law enforcement agency will work directly with the media. The utility should not communicate directly with the media unless the communications have been approved by the lead agency. If the event is a natural disaster or other noncriminal event, maintain the "speaking in one voice" concept.

When working with the media, there are some basic rules. Never release information that you are not prepared to see on television, hear on radio, or see in the newspaper seconds after it is released. Keep the media informed, and be honest. In a protracted situation when nothing new may happen for an extended period of time, have a media conference, and tell them nothing new has happened. When possible, provide handouts at media briefings.

The media provides an effective means of communicating with utility customers during a crisis. Learn to work with them and through them to reach your customers.

After the Crisis

The value of a lessons learned analysis cannot be overemphasized. Its primary purpose is to review your utility's response to the event and to improve on your plans and responses as needed. But don't keep this in-house. Lessons learned should be shared with other utilities through industry publications, presentations at meetings and conferences, and on the utility's web site. Even in a competitive environment, security experiences should be shared, especially since the lessons learned may have an impact on infrastructure protection and national security.

Initial responsibilities after the crisis include damage assessment and restoration of services. Activities that take place at this point include the identification of priorities and the resources needed to meet these needs. Inevitably "satisficing" will occur. *Satisficing* is a decision-making dynamic that refers to "settling for the less-than-optimal solution." A mutual-response utility may arrive on the scene with equipment that is different from what your system specifications require. Will you use their equipment to get the system operational or wait a few weeks until your equipment arrives? You use their equipment but try to record where and when you *satisfice*. You can return the system to your specifications later.

Provide "at-a-boys" to personnel who help during the crisis. Local personnel and employees from mutual response utilities left the Hurricane Hugo cleanup with t-shirts and hats that read, "I survived Hurricane Hugo." These were well-deserved badges of honor. The major utility in the state, Santee Cooper, also developed a comprehensive after-action report and video that it shared with other utilities and a public relations video that helped to provide closure and understanding to the consumers in the state.

The lessons learned analysis should include the following information:

- ***Review the actions the utility took at each of the four crisis management stages.*** What did you learn? How could the response have been improved? What did you do right?
- ***What if the situation had been worse?*** What additional problems would have occurred, and how would the utility have responded?
- ***Review all mutual aid agreements.*** Did the assistance you requested work as planned? What other assistance was needed?
- ***Share what you learned with other utilities.***

Summary

Most utilities are good at crisis management because, thanks to natural disasters, they get plenty of experience, but you must prepare for the worst-case scenario. Until several years ago, a utility in the southeast United States had never experienced an outage that affected more than 18% of its service area. One week before Christmas, they had an ice storm that disrupted 80% of their service!

An intentional attack on infrastructure will not be as random. The criminals responsible will plan to disrupt interdependent components of the system and to take advantage of the cascading effects.

From a security perspective, your emergency response plans must include procedures to respond to a major intentional event. If you don't think it can happen at your utility, watch again the news reports from the Oklahoma City bombing and the attacks on September 11, 2001. Most people thought those incidents could never have happened, but they did. Remember, too, the incidents that were prevented in Sacramento and Wise County, Texas. Domestic and international terrorists are targeting infrastructures. It may not be al-Qaida that attacks your system. It could be a disgruntled customer or ex-employee, an anti-government group, special interest extremist, or a mentally disturbed individual. Your utility is responsible for recognizing and addressing these threats. It must also be prepared to respond if an event occurs.

Utility security can never return to September 10, 2001. Welcome to the new paradigm.

Action Checklist

1. Review your emergency operations plan.
    a. Identify employees who are responsible for initial response to a natural disaster or security-related incident.
    b. Determine which law enforcement agency has jurisdiction if this is a criminal event.
    c. Ensure that the notification telephone numbers and other contact information are accurate.
    d. List mutual assistance contacts and procedures.
    e. Identify all personnel who may play a role in responding to an event.
    f. Exercise potential scenarios with the "real players."
    g. Plan for a long-term event. Use the FEMA, worst-case scenario approach.
2. Review the EOC.
    a. Establish primary and alternative centers at secure locations.
    b. Plan for long-term operation of the centers using personnel in 12-hour shifts.
    c. Determine situations when a full activation of the EOC is necessary and situations when a partial activation may be appropriate.
    d. Ensure that you are prepared to provide for the basic functions of the EOC.
    e. Assign personnel or teams of personnel to the personnel roles listed.
    f. Review your procedures for working with the media during a crisis. Make sure that all employees understand the "speaking in one voice" concept.
3. Review after the crisis.
    a. Develop a lessons learned analysis, and share it with others in the industry.
    b. Recognize when *satisficing* occurs, and be prepared to address these issues and procedures later.
    c. Provide "at-a-boys" to personnel who help in responding to the event.

Notes

1. Hooper, Michael K., "The Day the Earth Shook," *http://www.securitymanagement.com*

2. "Con Edison Completes Extraordinary Restoration Effort in Lower Manhattan," *Con Edison Media Relations*, September 19, 2001, *http://www.conedison.com*

3. Federal Emergency Management Agency, "Emergency Management Guide for Business and Industry," September 30, 2002, *http://www.fema.gov/library/bizindix.shtm*

Appendices

A

Threat Alert Levels and Physical Response Guidelines for the Electric Sector

A Model for Developing Organizational-Specific Threat Response Plans

Version 2.0

October 8, 2002

North American Electric Reliability Council

Goals

- Define threat alert levels for all alerts issued by the NERC Electricity Sector Information Sharing and Analysis Center (ES-ISAC) in cooperation with the NIPC or other government agencies. These threat alert levels and physical response guidelines, however, do not apply to facilities regulated by the Nuclear Regulatory Commission.

- Provide guideline examples of security measures that electric utility organizations may consider taking, based on the alerts issued.
- Ensure that the application of these electricity infrastructure alert levels are appropriate based upon the threat information received by the ES-ISAC from government sources, electricity sector participants, and other ISACs.
- Ensure threat information from the telecom, oil/gas, information technology, and other sectors is included, as appropriate, in the formulation of a threat alert.
- Note that threat alerts could be issued for a specific geographical area, such as "specific region only" or "specific city only," or by category, such as "specific type of facility."

Note:

- Your utility should not rely solely on the NERC or the NIPC for threat information or alert notifications. The utility must establish a liaison with local law enforcement and other appropriate agencies, and it is responsible for monitoring ongoing local threat information.
- In addition, the utility may initiate an appropriate threat alert on its own. Conditions that may warrant a local decision include natural disasters, labor problems, hazardous material incidents, or other incidents that impact the local community and/or the utility.

Threat alert level definitions

Threatcon ES-Physical-Green (Low) Threatcon ES-Physical-Green applies when no known threat exists of terrorist activity or only a general concern exists about criminal activity, such as vandalism, which warrants only routine security procedures. Any security measures applied should be maintainable indefinitely and without adverse impact to facility operations. This level is equivalent to normal daily operations.

Threatcon ES-Physical-Blue (Guarded) Threatcon ES-Physical-Blue applies when a general threat exists of terrorist or increased criminal activity with no specific threat directed against the electric industry. Additional security measures are recommended, and they should be maintainable for an indefinite period of time with minimum impact on normal facility operations.

Threatcon ES-Physical-Yellow (Elevated) Threatcon ES-Physical-Yellow applies when a general threat exists of terrorist or criminal activity directed against the electric industry. Implementation of additional security measures is expected. Such measures are anticipated to last for an indefinite period of time.

Threatcon ES-Physical-Orange (High) Threatcon ES-Physical-Orange applies when a credible threat exists of terrorist or criminal activity directed against the electric industry. Additional security measures have been implemented. Such measures may be anticipated to last for a defined period of time.

Threatcon ES-Physical-Red (Severe) Threatcon ES-Physical-Red applies when an incident occurs or credible intelligence information is received by the electric industry indicating a terrorist or criminal act against the electric industry is imminent or has occurred. This condition may apply as a result of an incident in North America outside of the electricity sector. Maximum security measures are necessary. Implementation of such measures could cause hardship on personnel and seriously impact facility business and security activities.

Physical response guidelines for threat alert levels

The following are examples of physical security measures to be considered for each threat alert level. These examples are not an exhaustive or all-inclusive list of possible security measures. The intent is to help define the scope for measures each organization may implement for its specific alert level response plans, based on its very specific requirements. Not all measures are applicable to all organizations. An organization may decide to reorder the sequence of some measures it deems appropriate to its environment and responsibilities. It also is expected that most organizations may need to develop additional, specific security measures. (The term *fail-over* refers to overcoming failures in the system.)

Threatcon ES-Physical-Green (Low)

1. Normal security operating standards and procedures.
2. Occasional workforce awareness messages or tabletop exercises, as appropriate.
3. All security, threat, and disaster recovery plans should be routinely reviewed and updated. An annual review is recommended as a minimum.

Threatcon ES-Physical-Blue (Guarded)

4. Workforce awareness messages to be alert for unusual activities and whom to report such activities.
5. Review operational plans and procedures and ensure they are up-to-date, to include:
   a. security, threat, disaster recovery, and fail-over plans.
   b. other operation plans as appropriate, *i.e.*, transmission control procedures.
   c. availability of additional security personnel.
   d. availability of medical emergency personnel.
   e. review all data and voice communications channels to assure operability, user familiarity, and backups function as designed.
   f. review fuel source requirements.

Threatcon ES-Physical-Yellow (Elevated)

6. Implement measures 1–5 if they have not already been implemented.
7. Ensure all gates, security doors, and security monitors are in working order and visitor, contractor, and employee access control are enforced.
8. Notify critical and on-call personnel.
9. Establish/assure communications with law enforcement agencies.
10. Identify additional business/site-specific measures as appropriate.

Threatcon ES-Physical-Orange (High)

11. Implement measures 1–10 if they have not already been implemented.
12. Review need to revise plans in measure 3, based on current intelligence, and include additional instructions as appropriate to the security/threat plans.
13. Place all critical and on-call personnel on alert, consider holding tabletop exercises.
14. Enforce safe zones around facilities per security plan.
15. Ensure all gates and security doors are locked and actively monitored either electronically or by "random walk-by procedures."
16. Implement enhanced screening procedures for:
    a. anyone entering the facility.
    b. all deliveries and packages.
17. Contact and coordinate with fuel suppliers, as necessary.
18. Inspect site fuel storage and HAZ-MAT (hazardous material) facilities.
19. Increase liaison with law enforcement, medical emergency services, and other entities.

20. Coordinate critical facilities security with neighbors:
    a. virtual neighbors such as other utility organizations.
    b. physical facility neighbors.
21. Consider emergency utility operations procedures appropriate to available threat intelligence.
22. Media releases should be reviewed with security/alert level coordinator prior to release.
23. Review plan for returning to Threatcon level Yellow, Blue, or Green status.
24. Additional business/site-specific measures as appropriate.

Threatcon ES-Physical-Red (Severe)

25. Implement measures 1–24 if they have not already been implemented.
26. Send nonessential personnel home, per business/site specific procedures.
27. Close the facility to all tours and visitors not related to the alert.
28. Consider having medical emergency personnel on-site, if possible.
29. Continuously monitor or otherwise secure all entrances and critical service facilities, such as substations, etc. This step may include use of armed security personnel.
30. Stop all mail and package deliveries directly to site.
31. Inspect all vehicles entering site.
32. Ensure all on-site personnel are fully briefed on emergency procedures.
33. Establish frequent communications with all appropriate law enforcement agencies for two-way updates on threat status.
34. Review plan for returning to Threatcon level Orange, Yellow, Blue, or Green status.
35. Additional business/site specific measures, as appropriate.

Additional Threatcon response considerations

Procedures to consider at each Threatcon level:

- ❑ Alert notification procedures
- ❑ Incident response
- ❑ Consequence management
- ❑ Distinguished visitor protection
- ❑ Operations security
- ❑ Access controls
- ❑ Barriers
- ❑ Lighting
- ❑ On-site security elements
- ❑ Information operations
- ❑ Response to weapons of mass destruction
- ❑ ______________________________
- ❑ ______________________________
- ❑ ______________________________
- ❑ ______________________________
- ❑ ______________________________
- ❑ ______________________________
- ❑ ______________________________
- ❑ ______________________________

Threatcon Response Action Set Matrix

The Threatcon response plan is not complete until you develop an action set matrix. The matrix outlines the specific action to be taken for each of the Threatcon responses and assigns responsibility for that action.

Each matrix includes the:

- Threatcon level
- numbered Threatcon response
- specific action required
- individual responsible for the specific action
- person to whom completion of the action is reported

Every response listed in the Threatcon response plan must be included in the matrix.

The following is an example of an action set matrix for Threatcon ES-Physical-Yellow, response measure 7.

Threatcon ES-Physical-Yellow

Measure 7

Ensure all gates, security doors, and security monitors are in working order and visitor, contractor, and employee access control are enforced.

1. ***Main entrance to facility:***
   1.1 Office manager will secure all external doors.
   1.2 Receptionist will secure lobby door to facility.
   1.3 Report actions to facilities manager.

2. ***Storage area:***
   - 2.1 Warehouse foreman secures all fence gates.
   - 2.2 Warehouse foreman secures door to warehouse.
   - 2.3 Report actions to security manager.

3. ***Security monitors:***
   - 3.1 Security manager inspects all monitors.
   - 3.2 Technician to inspect recording capability.
   - 3.3 Technician reports to security manager.
   - 3.4 Security manager reports to VP operations.

4. ***Visitor and contractor control:***
   - 4.1 Facility manager identifies all visitor and contractor personnel on site.
   - 4.2 Facility manager ensures that visitor badge controls are being used properly.
   - 4.3 Facility manager reports actions to security manager.

B

Threat Alert System and Cyber Response Guidelines for the Electricity Sector

Definitions of Cyber Threat Alert Levels

A Model for Developing Organization-Specific Cyber Threat Alert Level Response Plans

Version 2.0

October 8, 2002

North American Electric Reliability Council

Goals

- Define information systems and services (cyber) threat alert levels issued by the NERC ES-ISAC in cooperation with the NIPC or other government agencies. These alert levels and physical response guidelines, however, do not apply to facilities regulated by the Nuclear Regulatory Commission.

- Provide guideline examples of security measures that electric utility entities may consider taking, based on cyber alert levels issued.
- Ensure that the electricity infrastructure cyber threat alert levels are consistent with the threat information received by the NERC from government sources and other ISACs.
- Assure that threat information from the telecom, oil/gas, information technology, and other sectors is included as appropriate in the formulation of a cyber threat alert level.
- Note that cyber threat alert levels could be issued (for example) for a specific computer platform or a communications protocol or service, such as "Windows 2000" or "SCADA Communications."

Threat alert level definitions

Threatcon ES-Cyber-Green (Low) Threatcon ES-Cyber-Green condition applies when there is no known threat of cyber attack or only a general concern about hacker activity that warrants only routine security procedures. Any cyber security measures applied should be maintainable indefinitely and without adverse impact to business or expenses. This may be equivalent to normal daily conditions.

Threatcon ES-Cyber-Blue (Guarded) Threatcon ES-Cyber-Blue condition applies when there is a general threat of increased cyber (hacker intrusions, viruses, etc.) activity with no specific threat directed toward the electric industry. Additional cyber security measures may be necessary, and, if initiated, they should be maintainable for an indefinite period of time with minimum impact on normal business or expenses.

Threatcon ES-Cyber-Yellow (Elevated) Threatcon ES-Cyber-Yellow condition applies when a general threat exists of disruptive cyber activity directed against the electric industry. Implementation of additional cyber security measures is expected. Such measures are anticipated to last for an indefinite period of time.

Threatcon ES-Cyber-Orange (High) Threatcon ES-Cyber-Orange condition applies when a credible threat exists of disruptive cyber activity directed against the electric industry. Additional cyber security measures have been implemented. Business entities need to be aware that corporate resources will be required above and beyond those required for normal business or expenses.

Threatcon ES-Cyber-Red (Severe) Threatcon ES-Cyber-Red condition applies when an incident occurs or credible intelligence information is received by the electric industry indicating a disruptive cyber attack against the electric industry is imminent or has occurred. This condition may apply as a result of an incident in North America outside of the electricity sector. Maximum cyber security measures are necessary. Implementation of such measures could cause hardship on personnel and seriously impact facility business and security activities.

Cyber response guidelines for threat alert levels

Following are examples of security measures to be considered at each cyber threat level. This is not an exhaustive or all-inclusive list of possible security measures. The intent is to provide a scope of measures that each organization may implement for their specific threat response plan, based upon their own specific requirements. Not all measures are applicable to all organizations. Some organizations may decide to reorder the sequence of some measures, as they perceive appropriate to their environment and responsibilities. It is also expected that most organizations may perceive the need to develop additional, specific security measures to meet their requirements. It is also recognized that some measures might not always be necessary or applicable against a particular threat. Therefore, when developing your specific response plan, it is recommended you do so with consideration as a checklist of all the possible security measures you might choose to initiate, based on the specific threat information available.

Threatcon ES-Cyber-Green (Low)

1. Have an emergency plan for IT operations:
    a. Ensure all business critical information and information systems (including applications and databases) and their operational importance is identified.
    b. Ensure all points of access and their operational necessity are identified.
2. On a continuing basis, conduct normal security practices. For example:
    a. Conduct education and training for users, administrators, and management.
    b. Ensure an effective password management program is in place.
    c. Conduct periodic internal security reviews and external vulnerability assessments.
    d. Conduct normal auditing, review, and file back-up procedures.
    e. Ensure effective virus protection scanning processes are in place.
    f. Confirm the existence of newly identified vulnerabilities and test and install patches as available.
    g. Periodically review and test higher threat alert level actions and IT recovery plans.
3. Maintain law enforcement liaison [*e.g.*, local FBI, InfraGuard, Royal Canadian Mounted Police (RCMP)]

Threatcon ES-Cyber-Blue (Guarded)

4. Implement measures 1–3 if not already implemented.
5. Communicate workforce awareness messages to be alert and who to report unusual cyber activities to.
6. Review security and operational plans and procedures and ensure they are up-to-date.

Threatcon ES-Cyber-Yellow (Elevated)

7. Implement measures 1–6 if not already implemented.
8. Increase level of auditing, review, and critical file back-up procedures.
9. Conduct internal security review on all critical systems.
10. Increase review of intrusion detection and firewall logs.
11. Conduct more frequent checks of cyber security communications for software vulnerability.
12. Identify additional business/site-specific measures, as appropriate.
13. Increase frequency of measure 3; include additional instructions, as appropriate, to your cyber alert level response plan.

Threatcon ES-Cyber-Orange (High)

14. Implement measures 1–13 if not already implemented.
15. Conduct immediate internal security review on all critical systems.
16. Determine staffing availability for back-up operations, and provide notice.
17. Consider increasing physical access restrictions to computer rooms, communications closets, and critical operations areas.
18. Consider account access restrictions; temporarily disable noncritical accounts.
19. Consider delaying scheduled, routine maintenance, or nonsecurity sensitive upgrades.
20. Review media releases with cyber alert level coordinator prior to release.
21. Review plan for returning to alert advisory level Yellow, Blue, or Green.
22. Additional business/site-specific measures, as appropriate.

Threatcon ES-Cyber-Red (Severe)

23. Implement measures 1–22 if not already implemented.
24. Consider 7/24 emergency tech support staffing.
25. Consider continuous 7/24 monitoring of intrusion detection and firewall logs.
26. Consider continuous 7/24 monitoring of cyber security communications for latest vulnerability information. Contact software vendors for status of software patches and updates.
27. Consider reconfiguring information systems to minimize access points and increase security.
28. Consider rerouting mission-critical communications through unaffected system.
29. Consider disconnecting nonessential network access.
30. Consider alternative modes of communication, and disseminate new contact information, as appropriate.
31. Consider activation of the company emergency management team/procedures.
32. Actively monitor communications with all appropriate law enforcement and cyber security agencies for two-way updates on threat status.
33. Review plan for returning to advisory alert level Orange, Yellow, Blue, and Green.
34. Additional business/site-specific measures as appropriate.

C

Facility Vulnerability Determining System

The original *Facility Vulnerability Determining System* (FVDS) was developed in the 1970s for use by the U.S. Army and was first modified for nonmilitary use as Appendix C of the Antiterrorism Handbook. It was originally referred to as the "Installation Vulnerability Determining System."

No single factor should be a determinant in itself. The relationship between the factors and their relevance to your specific situation must also be considered. The system uses a scale of 0–100 points. The higher your score, the higher your potential vulnerability.

Avoid developing a "points mentality." Many facilities with a low point value are still primary targets for terrorists and other adversaries, whereas some facilities with a high point value may actually be low-priority targets.

The results of the FVDS can be used to assess your potential vulnerability to the general threats identified during your threat analysis. Each category of the FVDS includes space for you to consider the relationship between that category and the threats identified. By completing this section of the assessment, you will be better prepared to develop countermeasures to mitigate or manage your vulnerabilities.

The FVDS is usually completed by a small group of knowledgeable individuals, and the results then presented to the threat management committee.

1. Facility characteristics and sensitivity

18 Total Points

___ Very important persons located at this facility on a daily basis (6 points maximum):

- ❑ U.S. dignitary (1 point each)
- ❑ Foreign dignitary (3 points each)

___ Mission sensitivity (6 points maximum):

- ❑ Nuclear, chemical/natural gas/propane, law enforcement facility (6 points)
- ❑ Research and development (5 points)
- ❑ International corporation or activity (4 points)
- ❑ Domestic corporation (2 points)

___ Current threat analysis conducted by law enforcement or security professionals:

- ❑ Available and current (0 points)
- ❑ Unavailable or outdated (3 points)

___ Access control:

- ❑ Open access to facility (2 points)
- ❑ Limited access (1 point)
- ❑ Totally controlled access (0 points)

___ Symbolic value of facility: shrine, museum, *etc.* (1 point)

Comments All facilities should be capable of establishing barrier integrity and appropriate standoff parking distances, especially during increased threat situations.

Threat Considerations ______________________________

____

____

____

2. Status of training and security awareness briefings

12 Total Points

___ No operational EOC or immediate tactical law enforcement response capability (12 points)

___ Operational EOC but no immediate tactical law enforcement response capability (9 points)

___ Operational EOC, law enforcement response capability but no immediate tactical response capability (6 points)

___ Operational EOC, immediate tactical law enforcement response capability (3 points)

___ Operational EOC, immediate tactical law enforcement response capability. Tactical teams receive counter-terrorism at least every six months (0 points)

Comments Consideration should be given to establishing, equipping, maintaining, and testing of the EOC. Create a liaison with local law enforcement to determine the availability and training status of the tactical team response.

Threat Considerations ______________________________

____

____

____

3. Available emergency communications
10 Total Points

___ Emergency communications within organization only (4 points)

___ Emergency communications with law enforcement and other appropriate response or public safety agencies (3 points)

___ Redundant emergency communication capabilities with law enforcement and other appropriate response or public safety agencies (0 points)

___ Landline communications

- ❑ Nondedicated, dial-up (4 points)
- ❑ Dedicated point-to-point (2 points)
- ❑ Secure dedicated communication capability (0 points)

___ Digital/cellular/ radio communications

- ❑ Nondedicated (2 points)
- ❑ Dedicated (1 point)
- ❑ Secure and dedicated (0 points)

Comments Consideration should be given to security of communications including dedicated communication channels and the use of encryption. Secure communications are an important consideration for law enforcement, military, and security communications.

Threat Considerations ______________________________

____

____

____

____

____

____

____

4. Availability of law enforcement resources

8 Total Points

	Response Time			
	1 hr	*2 hrs*	*3 hrs*	*3 hrs+*
Trained federal and local	1	2	3	4
Trained federal	2	3	4	5
Trained local	3	4	5	6
Nontrained local	4	5	6	7
Unavailable	8	8	8	8

___ Points Assigned

Comments Consideration should be given to the determination of which law enforcement agencies have jurisdiction and are available to respond to an incident at this facility. Consider the resources, status, and training of each agency.

Threat Considerations ______________________________

____

____

____

____

____

____

____

____

____

____

____

____

5. Time and distance from other facilities or organizations with mutual response capability

7 Total Points

	Distance (Miles)			
Time (Hr)	**0–29**	**30–59**	**60–90**	**90+**
1 1/2	0	1	2	3
2	1	2	3	4
2 1/2	2	3	4	5
3	3	4	5	6
3+	4	5	6	7

___ Points Assigned

Comments Coordination should be made with the closest facility or organization capable of providing mutual assistance.

Threat Considerations ______________________________

____

____

____

____

____

____

____

____

____

____

____

6. Time and distance from urban areas

8 Total Points

	Distance (Miles)			
Time (Hr)	*0–59*	*60–89*	*90–120*	*120+*
1	8	7	6	5
2	7	6	5	4
3	6	5	4	3
4	5	4	3	2
4+	4	3	2	1

___ Points Assigned

Comments For purposes of this assessment, an urban area has a population of more than 100,000 people. In general, urban areas provide an opportunity for terrorists and other extremists to blend into the local population, and a safe haven from which to conduct their operations.

Threat Considerations ________________________________

____

____

____

____

____

____

____

____

____

____

7. Geographic Region

8 Total Points

___ New York City, Washington, D.C., West Coast, Florida, outside of the United States (8 points)

___ Eastern, including northeastern United States (6 points)

___ South and Southwest (4 points)

___ Northwest, Central, and Mid-Atlantic (2 points)

Comments Points are assigned based on historical data on terrorist activity by geographic region. This classification scheme should be modified as these trends change or as the terrorist activity in your region increases or decreases.

Threat Considerations ____

8. Population density of the facility

8 Total Points

Facility Population	Single Facility or Office	Small or Medium Office Building	Major Facility or Office Building
0–100	3	2	1
101–500	6	5	4
501–1,000	8	7	6
1,000+	8	8	8

___ Points Assigned

Comments If the facility is located in an office complex, industrial park, etc., consider the population density for the entire immediate area.

Threat Considerations ______________________________

____

____

____

____

____

____

____

____

____

____

____

____

9. Proximity to foreign borders

8 Total Points

___ Mexican border

- ❑ 0–100 miles (8 points)
- ❑ 101–500 miles (6 points)
- ❑ More than 500 miles (2 points)

___ Canadian border

- ❑ 0–100 miles (6 points)
- ❑ 101–500 miles (4 points)
- ❑ More than 500 miles (2 points)

Comments For facilities outside of the United States, assess the maximum point value.

Threat Considerations ______________________________

10. Access to the facility

8 Total Points

___ Roads

- ❑ Freeways or interstate highways (3 points)
- ❑ Improved roads (2 points)
- ❑ Secondary roads (1 point)

___ Airport or airfields

- ❑ Usable by high-performance (jet) aircraft (3 points)
- ❑ Usable by low-performance (prop) aircraft (2 points)
- ❑ Usable by small fixed-wing/rotary-wing aircraft (1 point)

___ Waterways

- ❑ Navigable (2 points)
- ❑ Nonnavigable (1 point)
- ❑ None (0 points)

Comments Consideration should be given to these three methods of entering or exiting that area around the facility from the points of view of the terrorist and also that of law enforcement who may be called to respond to an incident.

Threat Considerations ______________________________

____

____

____

____

____

____

____

11. Terrain around the facility

5 Total Points

___ Built-up areas (5 points)

___ Mountainous, forested, or areas conducive to concealment (4 points)

___ Open areas (2 points)

Comments Terrain should be analyzed in conjunction with a review of facility sensitivity, adequacy of barrier fencing, and routes of access and egress.

Threat Considerations ________________________________

____

____

____

____

____

Vulnerability determination

_____ Total points for all 11 categories

Point Total	Vulnerability
0–0	Very low vulnerability
11–30	Low vulnerability
31–60	Medium vulnerability
61–80	High vulnerability
81–100	Very high vulnerability

D

Computer and Internet Use Policies

Every employee and vendor who has access to your computers should read and sign the Computer Use Policy.

Every employee and vendor who has Internet access from your system should read and sign the Internet Use Policy.

Note: Some employees and vendors may have a need to use your computers and network but may not have a need for Internet access.

Computer Use Policy

[Name of Organization]

[Date and Policy Number if Appropriate]

Rights and responsibilities

Worldwide, open access electronic communication is a privilege, and continued access requires that users act responsibly. Users should be able to trust that the products of their intellectual efforts will be safe from violation, destruction, theft, or other abuse. Users sharing computing resources must respect and value the rights and privacy of others, respect the integrity of the systems and related physical resources, and observe all relevant laws, regulations, and contractual obligations. Users are responsible for refraining from acts that waste resources, prevent others from using them, harm resources or information, or abuse other people. To help protect files, users are responsible for setting passwords appropriately and for keeping passwords confidential by not giving them to another person.

Most [Name of Organization] owned computers are under the control of a system administrator or IT manager. These administrators are expected to respect the privacy of computer system users. However, [Name of Organization] computer system administrators may access user files or suspend services on the systems they manage without notice as required to protect the integrity of computer systems or to examine accounts that are suspected of unauthorized use, misuse, or have been corrupted or damaged. This includes temporarily locking vulnerable accounts, removing hung jobs, reprioritizing resource intensive jobs, etc. Some [Name of Organization] departments have their own computing and networking resources and policies. When accessing computing resources, users are responsible for obeying both the policies described here and the policies of those departments.

Examples of misuse

Examples of misuse include, but are not limited to:

- Knowingly running or installing on any computer system or network, or giving to another user, a program intended solely for the purpose of damaging or placing excessive load on a computer system or network. This includes, but is not limited to, computer viruses, Trojan horses, worms, flash programs, or password cracking programs.
- Attempting to circumvent data protection schemes or uncover security loopholes without prior written consent of the system administrator. This includes creating and/or running programs that are designed to identify security loopholes and/or intentionally decrypt secure data.
- Using computers or electronic mail to act abusively toward others or to provoke a violent reaction, such as stalking, acts of bigotry, threats of violence, or other hostile or intimidating "fighting words." Such words include those terms widely recognized to victimize or stigmatize individuals on the basis of race, ethnicity, religion, sex, sexual orientation, disability, and other protected characteristics.
- Posting on electronic bulletin boards or web pages materials that violate the organization's codes of conduct. This includes posting information that is slanderous or defamatory in nature or displaying graphically disturbing or sexually harassing images or text in a public computer facility or location that is in view of other individuals.
- Attempting to monitor or tamper with another user's electronic communications or reading, copying, changing, or deleting another user's files or software without the explicit agreement of the owner.
- Violating terms of applicable software licensing agreements or copyright laws.
- Using [Name of Organization] networks to gain, or attempt to gain, unauthorized access to any computer system.
- Using a computer account or obtaining a password without appropriate authorization.
- Facilitating or allowing use of a computer account and/or password by an unauthorized person.

- Masking the identity of an account or machine. This includes sending mail that appears to come from someone else.
- Performing an act without authorization that will interfere with the normal operation of computers, terminals, peripherals, networks, or will interfere with others' ability to make use of the resources.
- Using an account for any activity that is commercial in nature not related to work at [Name of Organization], such as consulting services, typing services, developing software for sale, advertising products, and/or other commercial enterprises for personal financial gain.
- Deliberately wasting computing resources, such as playing games, sending chain letters, spamming, treating printers like copy machines, storing or moving large files that could compromise system integrity or preclude other users' right of access to disk storage, etc.

Consequences of misuse

Misuse of computing, networking, or information is unacceptable, and users will be held accountable for their conduct. Serious infractions can result in temporary or permanent loss of computing and/or network privileges and/or federal or state legal prosecution. Appropriate corrective action or discipline may be taken in conformance with applicable personnel policies, collective bargaining agreements, and procedures established by federal and state codes that make certain abuses a crime, such as illegal reproduction of software protected by U. S. copyright law. Penalties can include a fine and/or imprisonment. Files may be subject to search under proper authorization.

Minor infractions of this policy, such as poorly chosen passwords, overloading systems, and excessive disk space consumption, are typically handled internally by the department in an informal manner. More serious infractions such as abusive behavior, account invasion or destruction, attempting to circumvent system security, etc. are handled formally using appropriate disciplinary procedures.

Contact information

Authorized users who have questions or comments regarding this policy should contact [Name of Contact Person].

Authorized users who suspect that these policies are being violated should contact [Name of Contact Person].

Internet Use Policy

[Name of Organization]

[Date and Policy Number if Appropriate]

Purpose

The Internet provides a source of information that can benefit every professional discipline represented by [Name of Organization]. It is our policy that employees whose job performance can be enhanced through use of the Internet be provided access and become proficient in its capabilities. This policy document delineates acceptable use of the Internet by employees, vendors, and contractors while using [Name of Organization] owned or leased equipment, facilities, Internet addresses, or domain names registered to [Name of Organization].

Background

The Internet is comprised of thousands of interconnected networks that provide digital pathways to millions of information sites. Because these networks subscribe to a common set of standards and protocols, users have worldwide access to Internet hosts and their associated applications and databases. Electronic search and retrieval tools permit users to gather information and data from a multitude of sources and to communicate with other Internet users who have related interests.

Access to the Internet provides [Name of Organization] with the opportunity to locate and use current and historical data from multiple sources worldwide in their decision-making processes. Employees and authorized vendors and contractors of [Name of Organization] are encouraged to develop the skills necessary to effectively utilize these tools in the performance of their jobs.

Scope of the policy

This policy applies to Internet access only. It does not cover the requirements, standards, and procedures for the development and implementation of information sites on the Internet.

The following Internet users are covered by this policy:

- Full or part-time employees of [Name of Organization]
- Vendors who are authorized to use these resources to access the Internet
- Contractors who are authorized to use [Name of Organization] owned equipment or facilities

This policy distinguishes between Internet access performed during normal working hours and that performed on personal time (*i.e.*, on weekends, before and after work, during lunch periods, or during scheduled break periods). This policy applies to Internet access when using [Name of Organization] equipment and facilities and performed using Internet Protocol addresses and domain names registered to our organization.

Policy

[Name of Organization] promotes Internet use that enables employees to perform work-related tasks and encourages its employees, vendors, and contractor personnel to develop Internet skills and knowledge. If an employee's supervisor determines that Internet access is in the best interest of [Name of Organization], the employee will be permitted, within the limits set forth later, to use the Internet on personal time to build his/her network search and retrieval skills. Employees who do not require access to the Internet as part of their official duties, may not access the Internet using these facilities under any circumstances. It is expected that employees will use the Internet to improve their job knowledge; to access information on topics that have relevance to their work; and to communicate with their peers in government agencies, academia, and industry. Users should be aware that when access is accomplished using Internet addresses and domain names registered to the [Name of Organization], they may be perceived by others to represent our organization. Users are advised not to use the Internet for any purpose that would reflect negatively on this organization or its employees.

[Name of Organization] computer systems are for work-related use and not for personal use; however, when certain criteria are met, authorized users are permitted to engage in the following activities:

- During working hours, access job-related information, as needed, to meet the requirements of their jobs.
- During working hours, participate in news groups, chat sessions, and e-mail discussion groups (list servers), provided these sessions have a direct relationship to the user's job. If personal opinions are expressed, a disclaimer should be included stating that this is not an official position of this organization.
- During personal time, retrieve non-job-related text and graphics information to develop or enhance Internet-related skills if the office pays a fixed rate for Internet access; *i.e.*, the access charge is usage insensitive, and, if a dial-up connection is made to an Internet access provider, it must be within the local calling areas. It is expected that these skills will be used to improve the accomplishment of job-related work assignments. By encouraging employees to explore the Internet, the [Name of Organization] also builds its pool of Internet-literate staff who can then guide and encourage other employees.
- Employees are prohibited from initiating non-work-related Internet sessions using [Name of Organization] information resources from remote locations, i.e., employees shall not dial into these resources

from home or other off-site locations for the purpose of participating in non-job-related Internet activities.

The following uses of the Internet (either during working hours or personal time) using [Name of Organization] equipment or facilities are not allowed:

- Accessing, retrieving, or printing text and graphics information that exceeds the bounds of generally accepted standards of good taste and ethics.
- Engaging in any unlawful activities or any other activities that would in any way bring discredit on our organization.
- Engaging in personal commercial activities on the Internet, including offering services or merchandise for sale or ordering services or merchandise from on-line vendors.
- Engaging in any activity that would compromise the security of any host computer. Host log-in passwords will not be disclosed or shared with other users.
- Engaging in any fund raising activity, endorsing any product or services, participating in any lobbying activity, or engaging in any active political activity.

Supervisory responsibility

[Name of Organization] employees, vendors, and contractors will have the final authority in determining whether an employee requires Internet skills to accomplish their assigned duties. Supervisors have the responsibility for:

- Acquiring Internet access for their employees who need it to conduct official business.
- Determining whether or not Internet access is provided to their employees for a flat fee. If access is provided on a flat fee basis, employees may use the Internet for the activities outlined above.

Supervisors should check with the group system or local area network (LAN) administrator to determine whether their Internet access is acquired at a flat fee.

- Advising their employees regarding the restriction against personal use of Internet access resources from other than [Name of Organization] facilities.
- Assuming the responsibility for making the final determination as to the appropriateness of their employees' use of the Internet when questions arise. This shall include the acceptability of Internet sites visited and the determination of personal time versus official work hours.

User responsibilities

Use of computer equipment and Internet access to accomplish job responsibilities will always have priority over personal use. In order to avoid capacity problems and to reduce the susceptibility of information technology resources to computer viruses, Internet users will comply with the following guidelines:

- Personal files obtained via the Internet may not be stored on individual PC hard drives or on LAN file servers.
- Official video and voice files should not be downloaded from the Internet except when they will be used to serve an approved job-related function.

Users are responsible for:

- Following existing security policies and procedures in their use of Internet services and will refrain from any practices that might jeopardize computer systems and data files (including, but not limited to, virus attacks) when downloading files from the Internet.
- Learning about Internet etiquette, customs, and courtesies, including those procedures and guidelines to be followed when using remote computer services and transferring files from other computers.

- Familiarizing themselves with any special requirements for accessing, protecting, and utilizing data, including Privacy Act materials, copyrighted materials, and procurement sensitive data.
- Conducting themselves in a way that reflects positively on the organization, since they are identified as [Name of Organization] employees on the Internet even though they may be using the Internet for personal reasons, as stated earlier.
- Being aware, along with their supervisors, whether Internet access is billed on a flat fee rather than a usage sensitive basis.

Individuals using [Name of Organization] equipment to access the Internet are subject to having activities monitored by system or security personnel. Use of this system constitutes consent to security monitoring, and employees should remember that most sessions are not private.

Contact information

Authorized users who have questions or comments regarding this policy should contact [Name of Contact Person].

Authorized users who suspect that these policies are being violated should contact [Name of Contact Person].

E

Personal Protection Checklist

Security at Home

- ❑ Make sure external doors have a sturdy dead-bolt lock with a minimum $1\frac{1}{2}$" bolt. Make sure all exterior doors are solid hardwood or metal. Install peepholes if needed.
- ❑ Secure sliding glass doors with commercial locks or with a broomstick in the track to jam the door.
- ❑ Secure double-hung windows with key locks, sliding bolts, or a nail in a hole drilled into the sash.
- ❑ Don't hide keys in mailboxes, under mats, or other obvious places. Give an extra key to a neighbor you trust.
- ❑ If you have moved into a new home or apartment, have the locks changed.
- ❑ Trim shrubbery around the residence, especially shrubbery that hides windows and doors.

- ❑ Use timers or motion detectors and outside security lights.
- ❑ Clearly display your house number so that police or emergency responders can locate you.
- ❑ Put inside lights and a radio or television on timers when you are away.
- ❑ If you arrive home and a door is open or a window broken, do not go into the house. Call the police from a neighbor's or on your cell phone.
- ❑ If you are at home and hear a noise that sounds like someone breaking in, quietly call the police, and wait until they arrive.
- ❑ Consider an alarm system. Use an established company. Check with the Better Business Bureau and your local police department for references. Learn how to use the system properly to avoid nuisance alarms.
- ❑ Keep a written record and photographs of all your valuables. Consider making a videotape of them in your home. Keep the list, photographs, and videotape in a safe deposit box at the bank.

Local Travel

- ❑ Make sure that your home looks occupied when you are not there.
- ❑ If you are away for several days or longer, stop your mail and newspapers, or have someone collect them for you.
- ❑ Leave shades and blinds in normal positions.
- ❑ Park a car in your driveway.
- ❑ If you are going to be gone for several days or longer, ask the police if the department has a "vacations check" program.
- ❑ Lock all doors and windows. Double check.

- ❑ Be alert to activity around or near your car.
- ❑ When approaching the car, have key in hand and check handles, locks, and back seat before entering.
- ❑ Keep the doors and windows closed and locked.
- ❑ When stopped in traffic, leave space between your car and the car in front of you.
- ❑ Be suspicious of people who approach you for directions or other information.
- ❑ If another driver bumps your car, drive to a congested area such as a shopping center or preferably to a police station. Call the police on your cell phone, and don't unlock or leave the car until they arrive. Tell the other driver, through the closed window and locked door, that you have called the police.
- ❑ Be alert when using a drive-up automated teller machine or post office box.

On the Road

- ❑ Provide an itinerary and contact numbers to a trusted person.
- ❑ Carry a minimum amount of cash. Use traveler's checks and credit cards, but keep a record of their numbers in a separate and safe place.
- ❑ Plan your route carefully. Use main roads and maps.
- ❑ Don't advertise your plans to strangers.
- ❑ Always lock your car when it is parked.
- ❑ Never pick up hitchhikers.
- ❑ Lock all packages and luggage in the trunk if possible.

- ❑ Always take your parking ticket with you. Leaving it in the car allows the thief to leave the parking lot unnoticed.
- ❑ When you stop overnight, remove all bags and other valuables from the car.
- ❑ Park in well-lighted areas.
- ❑ Carry a flashlight with fresh batteries, flares, a fire extinguisher, a first aid kit, and other emergency supplies.
- ❑ Avoid traveling during night hours if possible.
- ❑ Don't dress and act like a tourist.

At the Hotel

- ❑ Never leave luggage unattended.
- ❑ Use all available locking devices when in your room.
- ❑ Do not leave valuables in the room. Use the hotel's safe deposit box.
- ❑ Locate fire exits and other exits from the building.
- ❑ Never open the room door until you verify who is there. Use the peephole. If you have any doubts, call the front desk.
- ❑ Do not invite strangers into your room.
- ❑ If you see any suspicious activity, report it to the front desk.
- ❑ Ask hotel employees for specific directions when driving, and use the most direct route.
- ❑ If the family or group you are with goes to separate locations, agree first on a time and place to meet later.

Flying

- ❑ Dress casually in case you have to evacuate the airplane. Wear slacks or shorts. No tight fitting clothes. No skirts.
- ❑ Wear natural fibers. Synthetic clothing burns through to the skin.
- ❑ Do not wear high-heeled shoes.
- ❑ Drink plenty of water or juice before and during your flight.
- ❑ Carry medications with you. Do not put them in checked luggage.
- ❑ Get past the ticket counter to the secure area of the airport as quickly as possible.
- ❑ Once you have landed, proceed directly to baggage claim.
- ❑ Never agree to carry or to watch a package or luggage for a stranger.
- ❑ Report any unattended luggage.
- ❑ Leave the airport as quickly as possible.

Foreign Travel

- ❑ Protect your passport. It is your most important travel document and the primary target of many thieves.
- ❑ Carry valuables in closed pockets or in a waist belt under your shirt or coat.
- ❑ Before you leave, visit the U.S. Department of State website. Print the advisory for the countries you will be visiting and the locations and telephone numbers of the embassy and consulates in those countries (*http://travel.state.gov/*).

- ❑ Keep a low profile, and don't wear clothing that identifies your nationality.
- ❑ Avoid civil disturbances.
- ❑ Know and obey all local laws.
- ❑ Exchange money only at authorized exchanges.
- ❑ If you are driving overseas, learn the laws and customs before you leave. Drive carefully. Think twice before you decide to drive in major cities.
- ❑ Check on your medical and other insurance before leaving.
- ❑ Before you leave, get as much information as you can about your destinations. Visit the Internet, talk to your travel agent, visit the library, and buy a good guidebook.

F

Bomb Threat Checklist

Your name: ______________________________________

Date: ______ / ______ / ____________

Exact time call received: ______ : ______ ____ AM ____ PM

Call received at: (______) ______ – __________ ext __________

Questions to ask

- ❑ When is the bomb going to explode?
- ❑ Where is the bomb?
- ❑ What does it look like?
- ❑ What kind of bomb is it?
- ❑ What will cause it to explode?
- ❑ Did you place the bomb?
    - ❑ Why?

- ❑ Where are you calling from?
- ❑ What is your address?
- ❑ What is your name?
- ❑ Caller's voice (circle)

calm	slow	crying	slurred
stutter	deep	loud	broken
giggling	accent	angry	rapid
stressed	nasal	lisp	excited
disguised	sincere	squeaky	normal

- ❑ Other characteristics: ______________________________

Observations

- ❑ If the voice is familiar, whom did it sound like?

____

- ❑ Were there any background noises?

____

____

____

Other remarks or observations

Call reported to: ______________________________________

Date reported: ________ / ________ / ________________

Time reported: ________ : ________ ____ AM ____ PM

G

Executive Crisis Response File

A separate crisis response file should be maintained on each executive who is at risk. Files should be updated annually and should include photographs and videotapes of the executive's residence and office.

Executive Crisis Response File

Date: ______ / ______ / ____________

Name: ____

Position: ____

Information contained in file

	Names	*Profile*	*Photo*	*Finger prints*	*Writing samples*	*Family Voice tapes*
Executive	________________	___	___	___	___	___
Spouse	________________	___	___	___	___	___
Children	________________	___	___	___	___	___
	________________	___	___	___	___	___
	________________	___	___	___	___	___
	________________	___	___	___	___	___

Other family members
Relationship

________	________________	___	___	___	___	___
________	________________	___	___	___	___	___
________	________________	___	___	___	___	___
________	________________	___	___	___	___	___
________	________________	___	___	___	___	___
________	________________	___	___	___	___	___

	Sketches	*Photographs*
Residence:	___	___
Second residence:	___	___
Vehicle #1		___
Vehicle #2		___

Individual profile

Date Completed: _____ / _____ / ___________

Name: ____

Last *First* *Middle*

Nickname(s): ____

Social Security Number: _______ – _______ – __________

Driver's License: ____

State *Number*

Birthdate: _____ / _____ / ___________

Birthplace: ____

City *State/Province* *Country*

Address: ____

____

____

City *State/Province* *Postal Code*

Physical description:

Height: __________ Weight: _____________

Hair: ___________ Eyes: _______________

Eyeglasses: _______ Hearing aid: __________

Other: ________________________________

Scars or identifying marks:

____

____

____

Identification records

ATTACH
PHOTO
HERE

ATTACH
PHOTO
HERE

ATTACH
FINGERPRINT
RECORD
HERE

Medical profile

Physician: Name: ______________________________

Address: ______________________________

____

Telephone: (______) ______ – ____________

Dentist: Name: ______________________________

Address: ______________________________

____

Telephone: (______) ______ – ____________

Pharmacist: Name: ______________________________

Address: ______________________________

____

Telephone: (______) ______ – ____________

Medications: ______________________________

____

____

Social profile

Clubs/Organizations: ______________________________

____

____

Hobbies: ______________________________

____

____

Other Activities: ______________________________

____

____

Routine Events: Sunday: ______________________________

____

Monday: ______________________________

____

Tuesday: ______________________________

____

Wednesday: ______________________________

____

Thursday: ______________________________

____

Friday: ______________________________

____

Saturday: ______________________________

____

Financial profile

Bank(s): Name: ______________________________

Address: ______________________________

____

Telephone: (_______) _______ – ___________

Notes: ______________________________

____

Name: ______________________________

Address: ______________________________

____

Telephone: (_______) _______ – ___________

Notes: ______________________________

____

Broker(s): Name: ______________________________

Address: ______________________________

____

Telephone: (______) ______ – __________

Notes: ______________________________

____

Name: ______________________________

Address: ______________________________

____

Telephone: (______) ______ – __________

Notes: ______________________________

____

Credit Cards: Company: ______________________________

Address: ______________________________

____

Telephone: (______) ______ – __________

Acct #: ______________________________

Company: ______________________________

Address: ______________________________

____

Telephone: (______) ______ – __________

Acct #: ______________________________

Vehicle profile

Include boats, campers, and other recreational vehicles:

Make/Model	***Year***	***Color***	***License #***	***Vehicle ID #***

Residence Profile

Include sketched floorplans and photographs of both primary and additional residences, as well as any vacation properties:

ATTACH
PHOTO
HERE

Address: ______________________

____

____

City ***State/Province*** ***Postal Code***

ATTACH
PHOTO
HERE

Address: ______________________

____

____

City ***State/Province*** ***Postal Code***

Emergency contacts

Relatives not living with employee:

Relationship	*Name*	*Address*	*Telephone*

Children:

Name	*Grade*	*School*	*Address*	*Telephone*

Closest friends:

Name	*Address*	*Telephone*

Writing samples

Please write the following and sign your name using your regular signature:

"I feel good and look forward to tomorrow."

Signature *Date*

Printed Name

Signature *Date*

Printed Name

Signature *Date*

Printed Name

Family voice tape sequence

Voice Number	*Name*	*Relationship*
______	________________________________	__________________
______	________________________________	__________________
______	________________________________	__________________
______	________________________________	__________________
______	________________________________	__________________
______	________________________________	__________________
______	________________________________	__________________
______	________________________________	__________________

Sample Theft of Service Policy

Theft of Service Policy

i

ii

I. Most common violations and procedures for handling these situations

(A) Jumpers on the rear of the meter or in the meter socket

(B) Illegal connection at the weather head

(C) Illegal connection in an inactive account with reconnect order if more than 100 kwh has been registered since the disconnect

(D) Illegal reconnect on a nonpay account

(E) Illegal reconnect on a nonpay account with a reconnect order

(F) Meter bypass

(G) Meter upside down

(H) Damaged or removed seal, lock, or locking band

(I) Damaged meter found at location that has been disconnected for nonpayment, utility theft, returned check, or service disconnect

(J) Illegal connection on inactive account

(K) Use of stolen meter or any other unauthorized meter at an inactive or active account

(L) Fraudulent use of name on account to obtain service

Meter readers and service personnel: Upon discovering any of the above conditions, follow the procedure described below.

(A) If the situation is a "suspect" theft or fraud, (*e.g.*, a cut seal) report the situation to your supervisor or to the codes enforcement officer at the end of the workday.

– 1 –

(B) If the situation is obvious tampering (*e.g.*, meter inverted), report to the dispatcher immediately, and continue reading meters or working work orders.

(C) If the situation is dangerous (*e.g.*, an open meter socket), report to the dispatcher immediately, and wait at the location until a service person arrives to render the service safe. If the customer is present, tell the customer there is an unsafe condition and a crew is en route to correct it.

Dispatcher: When informed by meters readers or other field personnel of an obvious theft or an unsafe condition, notify the codes enforcement officer and appropriate service personnel immediately.

II. Illegal reconnect on an inactive account with reconnect order when less than 100 kwh has been registered since the last disconnect

Service personnel: Upon discovering usage of less than 100 kwh on an inactive account that has a connect or reconnect order, record the meter reading and the findings on your work order, and connect the service unless there is damage or a yellow seal on the meter. If you find damage or a yellow seal, do not connect or reconnect the service.

III. Cut seals

Service personnel: Report all cut or missing seals to the dispatcher.

Dispatcher: Forward all reports of cut or missing seals to the codes enforcement officer.

– 2 –

IV. Extension cords

When extension cords are used to provide power to a location without service, the service at the active account will be terminated immediately. Both customers must contact the codes enforcement officer before service can be reconnected at either account. The use of extension cords in these cases creates a potentially dangerous situation that could result in fire and loss of life.

Service personnel: Upon finding extension cords used to provide service to an inactive account, immediately report the condition to the dispatcher. Remain at the location until the service at both locations is disconnected.

Dispatcher: When notified of extension cords being used to provide service to an inactive account, immediately dispatch a crew to disconnect both services, and notify the codes enforcement officer.

V. Damaged meter

A. Trouble call:

Service personnel: Upon finding a damaged meter when dispatched to a location because of a reported service problem, contact the dispatcher before correcting the situation and restoring service.

Dispatcher: Upon receiving a call from the service person, ensure that the service is not disconnected as a result of nonpayment, utility theft, or for other reasons and that it is an active account. Unless one of these

– 3 –

reasons exists, instruct the service person to correct the situation and restore service.

B. Meter accidentally damaged by service personnel:

Service personnel: If the meter is damaged by personnel working on the service, restore the service and make appropriate meter change orders. Record what occurred. No disciplinary action will be taken, but there may be a need to review the use of equipment, the procedures followed, or the need for additional training.

VI. Damaged or removed locking band or lock, not resulting in illegal use of power

Service personnel: Report the situation to the dispatcher immediately. Leave the situation as found. If the situation is not dangerous, continue reading meters or working work orders.

Dispatcher: Report immediately to codes enforcement officer.

VII. Yellow seals

Service personnel: Yellow seals are used only on accounts where tampering has been detected or where it is suspected. Yellow seals should not be removed under any circumstance unless authorized by the codes enforcement officer, the operations manager, or the general manager. If a yellow seal is found cut or otherwise damaged, notify the dispatcher immediately.

– 4 –

Dispatcher: When receiving a report of a cut or damaged yellow seal, notify the codes enforcement officer immediately.

VIII. Fraudulent service

Some customers attempt to obtain fraudulent service by using incorrect names on the service application. This is especially prevalent where service has recently been disconnected for nonpayment.

Field personnel: If you receive a connect order for a service that you believe may be fraudulent, do not connect the service. Notify the dispatcher of your suspicion and the reasons for being suspicious.

Dispatcher: Upon receiving the report from the field personnel, notify the customer service manager and the codes enforcement officer. The customer service manager will review the records on the account to determine if fraud may be occurring. The codes enforcement officer will inspect the service. Service should not be connected until approved by the customer service manager.

Customer service manager: If it is determined that fraud may be occurring at the account, the customer will be required to come to the service center to verify his/her identity and to personally attest that a disconnected customer is not living at the residence. If fraud is occurring or has been attempted, the customer will be required to pay all back bills and current charges before service is

– 5 –

connected. If it cannot be determined that fraud is occurring or has been attempted, the service will be connected.

IX. Danger to life, limb, or property

Under no circumstances are these procedures intended to override or supersede the obligation of [Name of Organization] to prevent, if possible, the injury or loss of life, limb, or property. Safety should always be the first consideration.

X. Best judgment

These procedures cannot cover all of the situations involving tampering and fraud that employees find in the field. These procedures provide guidelines that should be used in combination with the employee's knowledge and experience. When in doubt as to how to respond to a unique situation, contact the dispatcher who will then contact the appropriate personnel to assist you.

– 6 –

I

Physical Security Plan

1. **Purpose** State the purpose of your plan.
2. **Area security** Define the areas, buildings, and other structures considered critical, and establish priorities for their protection.
3. **Control measures** Define and establish restrictions on access and movement into critical areas. These restrictions can be categorized as personnel, vehicles, and material assets.

    A. Personnel (employee, visitor, vendor) access

    1. Controls pertinent to each area or structure

        a. Authority for access

        b. Access criteria for

        (1) Employees

        (2) Visitors

        (3) Vendors

        (4) Maintenance persons

        (5) Contractor employees

        (6) Others

2. Identification and control
    a. Describe the system to be used in each area, *e.g.*, if a badge system is to be used, who issues the badge, what type of badge, how often will the badges be changed, *etc*.
    b. Application of the system to include all of the categories listed in A.1.b.

B. Material control
1. Incoming
    a. Requirements for admission of material and supplies
    b. Search for hazards and accountability of materials
    c. Special controls over delivery of supplies or other shipments into restricted areas
2. Outgoing
    a. Documentation required
    b. Controls
    c. Sensitive shipments

C. Vehicle control
1. Policy on company vehicles
2. Policy on employee vehicles
3. Policy on vendor and visitor vehicles
4. Controls for entry into restricted areas for each of the above plus emergency vehicles
5. Vehicle registration policy and procedures

4. Aids to security Indicate the manner in which the following listed aids to security will be implemented within the company areas.

A. Protective barriers
1. Definition
2. Clear zones
    a. Criteria
    b. Maintenance

3. Signs
    a. Types
    b. Posting
4. Gates
    a. Hours when open and when closed
    b. Security requirements
    c. Lock security

B. Protective lighting system
1. Use and control
2. Inspection (should be done during darkness)
3. Action to be taken in the event of a power failure
4. Action to be taken if alternative power fails
5. Emergency lighting systems
    a. Stationary
    b. Portable

C. Intrusion detection systems
1. Security classification
2. Inspection and testing
3. Use and monitoring
4. Action to be taken if alarm is triggered
5. Maintenance
6. Alarm logs or registers
7. Sensitivity settings
8. Fail-safe or tamperproof provisions
9. Monitor panel location

D. Communications
1. Locations
2. Specific equipment and backup equipment
3. Use
4. Tests

5. Authentication
    a. Frequencies
    b. Codes

5. **Security forces** Include general instructions that would apply to all security force officers (fixed and mobile). Detailed instructions, such as policy and procedures, need attachment as appendices.
    A. Composition and organization
    B. Length of duty shifts
    C. Essential posts and routes
    D. Weapons and equipment
    E. Training
    F. Use of sentry dogs
    G. Liaison with law enforcement

6. **Contingency plans** Indicate required actions in various emergency situations. Detailed plans, such as those for handling bomb threats, disasters, fire, *etc.*, need attachment as appendices.
    A. Individual actions
    B. Alert or reserve force actions
    C. Security force actions
    D. Increased security measures to be implemented at threat levels 2 and 3
    E. Coordination with outside agencies
        1. Police
        2. Fire
        3. Federal agencies

Appendices

A. Intelligence collection
B. Company security status map
C. Contingency plans
D. Special instructions to security officers
E. Other

J

Security Survey Checklist

Location: ____

Date: ______ / ______ / __________

Exterior Security

- ❑ External openings are properly secured.
- ❑ Doors are constructed of sturdy materials or reinforced.
- ❑ Entrances are equipped with secure locking devices.
- ❑ Doors are locked when not in use.
- ❑ Hinge pins to all entrance doors are spot welded or otherwise protected.

- ❑ Ventilators/other openings are covered by steel bars or mesh.
- ❑ Windows are securely fastened from the inside.
- ❑ Window openings less than 18 feet above ground are protected by steel bars/mesh.
- ❑ Clear areas are maintained around fences and buildings.
- ❑ Public parking is away from buildings.
- ❑ Access to utility and employee parking is controlled.
- ❑ Assigned parking spaces do not have names on them.
- ❑ Vehicles entering area are controlled.
- ❑ Exterior of building is inspected for signs of intrusion weekly.

Key Control

- ❑ Key control security officer has been assigned.
- ❑ Keys are issued only to authorized employees.
- ❑ Keys not in use are stored in a fireproof cabinet.
- ❑ Records are maintained on all keys issued.
- ❑ Records are maintained for all master keys.
- ❑ Locks are changed immediately when keys are missing or when people leave the utility.
- ❑ Locks are checked periodically by supervisor or manager.
- ❑ All rooms are locked when not in use.

Intrusion Detection System

- ❑ Alarm system has been properly installed.
- ❑ System is backed up by security officer or law enforcement personnel.
- ❑ System is inspected and checked regularly.
- ❑ System has backup power supply.
- ❑ System is tamper resistant and weatherproof.
- ❑ Records are kept of all alarm signals.

Protective Lighting

- ❑ Protective lighting is provided during hours of darkness.
- ❑ Interior emergency lighting is tested monthly.
- ❑ Inoperative lights are repaired or replaced immediately.
- ❑ Security check is conducted during darkness on a regular schedule.
- ❑ Security lighting log is maintained, and replacement is scheduled at 80% of the expected life of the lamp.

Access control to building or area

____

____

____

____

____

Material handling procedures

Cash handling procedures

Employee personal protection

Comments

Helpful Web Site Addresses

Computer Emergency Response Team
http://www.cert.org

Computerworld *(publication)*
http://computerworld.com

National Security Agency
http://www.nsa.gov/isso/index.html

U.S. Department of State
http://travel.state.gov

North American Electric Reliability Council
http://www.nerc.com

NIPC Watch and Warning Unit
http://www.nipc.gov

Trend Micro Weekly Virus Report
www.trendmicro.com/en/security/report/overview.htm

Video for Mail Handlers
www.osha-safety-training.net/ANT/anthrax.html

Center for Disease Control
www.cdc.gov

Computer Emergency Response Team (CERT) at Carnegie Mellon Software Engineering Institute
www.cert.org

CERT summaries
www.cert.org/summaries

Demographic statistics
www.fedstats.gov

Major search engine
www.google.com

Interagency OPSEC support staff
www.ioss.gov

International Utility Revenue Protection Association
www.iurpa.org

National Counterintelligence Executive
www.ncix.gov

CyberNotes publication
www.nipc.gov/cybernotes/cybernotes.htm

Bureau of Justice statistics
www.ojp.usdoj.gov/bjs

Southern Poverty Law Center
www.splcenter.org

The Hacktivist
www.thehacktivist.com

U. S. Department of Justice

www.usdoj.gov

Mail Center Security Guide

www.usps.com/cpim/ftp/pubs/pub166/welcome.htm

YellowPages.Com, Inc.

www.yellowpages.com

Index

A

B

C

D

E

F

G

H

I–J

K

L

M

N

O

P–Q

R

S

T

U

V

W–Z